INFINITE
NATURE

INFINITE

NATURE,

R. BRUCE HULL

The University of Chicago Press CHICAGO & LONDON

R. BRUCE HULL is professor of natural resources at Virginia Tech and coeditor of *Restoring Nature: Perspectives from the Social Sciences and Humanities.*

The University of Chicago Press, Chicago 60637
The University of Chicago Press, Ltd., London
© 2006 by The University of Chicago
All rights reserved. Published 2006
Printed in the United States of America

15 14 13 12 11 10 09 08 07 06   1 2 3 4 5

ISBN: 0-226-35944-1 (cloth)

Library of Congress Cataloging-in-Publication Data

Hull, R. Bruce.
Infinite nature / R. Bruce Hull.
p. cm.
ISBN 0-226-35944-1 (cloth : alk. paper)
1. Philosophy of nature.   2. Ecology—Philosophy.   I. Title.
BD581.H85   2006
113—dc22

2005035108

♾ The paper used in this publication meets the minimum requirements
of the American National Standard for Information
Sciences—Permanence of Paper for Printed Library Materials, ANSI Z39.48-1992.

To my other loves: Elizabeth, Quinn, and Constance

Nurturing, providing, mothering
Competitive, indifferent, risky

Resources, jobs, wealth
Life, habitat, services

Pollen, flower, garden,
Wolf, pack, ecosystem

Human, ape, evolution
Created, perfect, God

Static, resilient, balanced
Dynamic, frail, disturbed

Known, measured, managed
Mysterious, chaotic, unpredictable

Pure, clean, healthy
Hurricane, fire, infested

Restorative, exciting, fun
Pest, plague, death

Wild, untamed, uncooked
Domesticated, feral, engineered

Remote, isolated, protected
Potted, planted, managed

Creative, persistent, other
Child, dependent, lover

Many natures
Which should exist?

Contents

Preface

I'm a rebound romantic. I once was blindly in love with nature—romantically and lustfully in love. I immersed myself in the deep waters and penetrating solitude of wilderness lakes at the border between Minnesota and Canada, bushwhacked through dense rhododendron forests up steep Appalachian Mountains in pursuit of sunsets, and lost hours meditating on changing colors deep within the Grand Canyon. I opened windows to let flies escape and did not fish for sport because I found no pleasure in causing fish pain. I followed a vegetarian diet, took short showers, turned off lights, and faithfully recycled all materials my community allowed. I tried to take only memories and leave only footprints. But I was painfully aware that by living my life, I created trails in wilderness, trash in dumps, and carbon in the atmosphere. I felt guilty about being human and destroying the nature I loved.

The guilt grew as I took advantage of life's opportunities to travel the world and raise a family. I had to ignore or suppress this guilt in order to stay sane while leading a professional, middle-class American lifestyle. I knowingly became a hypocrite. I stocked bathrooms in my old but sturdy wood house with ample supplies of soft, clean toilet paper. I drove a car to work more often than I biked. I drank imported beer in nonreturnable bottles. I started eating steak again, and I liked it. On cold snowy days, I appreciated the warmth and comfort of my oil-fired furnace. I flew to conferences, drove on holidays to visit family members, and surfed the Web.

As I struggled with being a hypocrite, it soon dawned on me that I also am a bigamist. I love both nature and culture. I'd been captivated by fine music and literature, mesmerized by Michelangelo's *David*, and awed by the Great Wall of China. I love sipping fine wine at Australian vineyards and strolling dreamily through the tended Tuscan landscapes. I eagerly enrich my life with friends and ideas from around the world and marvel at the beauty and integrity humans create.

The first part of my professional career focused on the study, management, and protection of romantic experiences in nature. I taught in landscape architecture and natural resource recreation programs, conducted studies in natural areas, and developed methods to assess and legitimize public preferences for nature. Then, with the pressures of establishing a career and family behind me, and awareness of my hypocrisy and bigamy in front of me, I sought to resolve some contradictions in my life.

As a good romantic, I looked to nature for lessons. Pillars of U.S. environmentalism such as Thoreau, Leopold, and Muir found inspiration and guidance in nature, so I searched for answers in wild places and in the scientific studies of nature. I was disappointed to find no absolutes, only qualifications. Rather than an inspiring nature that knew best and could help me establish priorities and defend values, I learned about a dynamic and capricious nature. I found deep divisions and intense debate among my natural science colleagues over issues I had taken for granted: biodiversity is good, ecosystems have integrity, forests have health, species are entities, and naturalness can be defined. And I confronted the dark side to eco-philosophy that promotes fascism, social Darwinism, and misanthropy. Where Muir found humility and interconnectedness, Hitler found genocide and brutality.

I felt like my lover had abandoned me. If nature really was dynamic, capricious, and arbitrary, then perhaps I could be too. My guilt began to ease and my embrace of things cultural grew; clearly I was on the rebound from a shattered relationship and needed to be careful of new infatuations. I remained uneasy about my inability to judge environmental policies and evaluate alternative development scenarios. In what environment did I want to live? What environmental qualities should I advocate? How could I defend my preferences? I had so many more questions than answers that I soon felt more helpless than guilty. My most troubling realization was that I could not define or defend the nature I had loved. What was my lover? A fantasy? An expectation? A social construction?

My tools and training failed to help me make sense of these questions, so I redirected my professional career and began studying environmental debates and land-use planning efforts. I wanted to learn how others were resolving life's contradictions. Obviously many people had strong opinions (as evidenced by negotiation train wrecks, endless litigation of land-use plans, and heated scholarly debates about the nature of nature). I interviewed scores of people—some experts, some not—about their environmental preferences and concerns. I found that most people were at least as confused as I, and often became visibly anxious when my questions penetrated through their deep-seated beliefs and assumptions about nature and about humanity's relationship to nature.

Why is nature so elusive a lover? Why did I struggle defining it, justifying its existence, explaining my values for it, and figuring out how I should relate to it? Why do we as a society have such a hard time finding agreement on essential questions about the environmental conditions we want for ourselves and our descendants? The answer, I now suspect, is because environmental fundamentalism traps us in narrow, self-reinforcing, and polarizing debates. Actively pluralizing nature provided me the means to overcome these fundamentalist traps. Hope of deliberation and discourse replaced the helplessness of polarization and paralysis. Optimism about finding common ground replaced the negativity of pointing out differences. I rebounded back into a relationship with many natures. This book describes my journey toward pluralizing nature. I wrote it as a means to help understand and inventory some natures that can, have, and might exist.

The book's intended audience is everyone who cares about living in thriving and sustainable communities. That should include you. It certainly includes environmental professionals, environmental scientists, environmentalists, community leaders, politicians, parents, and anyone else involved with or responsible for determining the environmental qualities of our future. The book is meant to be accessible to a general audience. It makes no assumptions about prior exposure to evolution, ecology, economics, environmental history, religious philosophy, or ethical theory—but it combines and focuses these and other disciplinary perspectives on contemporary environmental issues. I've used the text as part of a general education course available to freshmen in every discipline and profession. The class and text evoke spirited discussion as well as frustration and joy at the complexity of our environmental challenges.

The book has been reviewed and improved by too many to mention, but I'd like to specifically acknowledge Paul Angermeier, Brian Britt, Terry Daniel, Erin DeWitt, Christie Henry, Brian Nettleton, David Robertson, Joe Roggenbuck, Gyorgyi Voros, as well as the hundreds of students who worked through early drafts.

Introduction

Environmental Fundamentalism
Unifying Visions of Thoreau and Leopold
Pluralizing Nature

Critics say environmental*ism* is dying; or if it is not dying, then it needs to. Traditional critics repeat well-worn complaints that environmentalism's agenda is misplaced, oppressive, and misanthropic because it produces regulatory bungles, slows economic growth, and delays technological advances that save lives. A newer line of criticism, coming from inside the movement, argues that environmentalism's alarmist rhetoric, polarizing ideology, and preservationist solutions are outdated and insufficient to the tasks of sustaining thriving communities in a humanized biosphere.[1] All these critics are wrong in one important regard: there is not one environmentalism, one environmentalist position, or one environment. Critics know this to be true but still struggle to escape the trap of environmental fundamentalism.

The trap of environmental fundamentalism gets sprung early and often. Environmental issues tend to get framed as either-or, win-lose debates: economy *or* environment, humans *or* nature, government regulation *or* market economy, preservation *or* development, growth *or* steady state. Serious public dialogue about desirable future conditions quickly polarizes and degenerates into name-calling, as it did in a recent planning effort for the lands within the Adirondacks' famous Blue Line. Environmentalists were stereotyped as "forest faggots," "nature Nazis," and "watermelons": green on the outside, red (socialist) on the inside. Non-environmentalists were typecast as rapists, destroyers, and greedy exploiters with warped priorities that value the almighty dollar above the Almighty's creation.[2]

Alternative framing of society's social-environmental problems is possible. Biocultural visionaries advocate appropriate technology and social ecology that blend rather than separate environmental and social concerns. They try to advance solutions that benefit both the environment *and* jobs, human equity *and* biodiversity, urbanization *and* ecology, utility *and* beauty, and thriving *and* sustainable communities. These attractive visions of the future appeal to people of most political persuasions, broadening and deepening the political will to act. To identify and realize these visions, we need to overcome the polarization caused by environmental fundamentalism. The purpose of this book is to assist in this effort by pluralizing our conception of nature.

Early and foundational environmental thinkers such as Henry David Thoreau and Aldo Leopold were unifiers, not polarizers.[3] They were highly critical of economics as the sole criterion for justifying land-use decisions and social policies. However, they were not anti-economic. Thoreau, for example, went to the woods to live deliberately, and his experiment at pond's edge was, in part, economic. The first and longest chapter of *Walden* is titled "Economy," in which he carefully records expenses and profits down to the half cent. He describes building a house, planting and hoeing a bean field, selling excess beans, and purchasing supplies with the profits. He rejoices in the fulfillment earned through the labor of living frugally, simply, deliberately, and economically.

However, Thoreau also soundly criticized his neighbors for overemphasizing economic values. Their narrow-minded, economic-only accounting of life trapped them in a sad, desperate cycle of toiling at jobs in order to secure the funds needed to purchase the products that others toiled to produce. His neighbors seemed so focused on staying atop the economic treadmill that they forgot to smell the roses and taste the fruits of life. Thoreau was not against labor but thought people sacrificed essential spheres of life—such as self-actualization, community, place, and posterity—by too eagerly selling their humanity as labor. He encouraged people to drive life into a corner and find in it deep meaning, spiritual fulfillment, aesthetic pleasure, ecological literacy, and community vitality. His two-year experiment at pond's edge was an effort to document different ways to understand and value living a simple life near nature.

That was 150 years ago. Today Thoreau might conclude that things have gotten worse. He also might conclude that the environmental movement he unknowingly helped found has failed to offer society a viable alternative. But he might see signs for hope that his agenda will be realized. While writing this introduction, I traveled to Walden Pond in search of Henry's muse. Urban sprawl long ago eroded much of the solitude Thoreau found during his long walks around Concord, Massachusetts. The edge of the famous pond is now badly beaten and bruised by the hundreds of thousands of visitors who frolic there each year. The pond regularly fills with boaters, swimmers, dogs, hikers, deer, and pilgrims such as me.

Extensive efforts are under way to restore functioning ecological systems to the pond's trampled watershed. Steel-wire fences and strongly worded signs direct visitors to narrow trails that ring the pond. Vegetation is being planted by volunteers, and soil has been secured by high-tech landscaping fabric. It is ironic that Thoreau's nightmare of a nature completely defined, dominated, and denuded by humanity occurred at the very place where he wrote some of his most powerful prose warning us of the dangers of an unrestrained instrumental worldview. It is perhaps fitting that restoration efforts are under way to balance the needs of a thriving, creative ecological system with the needs of a thriving, creative human civilization. Perhaps neither nature nor humanity will dominate this new arrangement and both will be better off for the partnership.

Aldo Leopold also was a unifier, appreciating both the economic and cultural harvests of nature. He studied, advised, and advocated hunting, agriculture, and other consumptive economic practices, recognizing that they provide the needed material harvests of food, shelter, comfort, and safety. He strove to harmonize economic practices with the integrity of functioning ecological systems and thus was critical of land management driven exclusively for economic gain. The land, he argued, also generated cultural harvests, providing people with a life as well as a livelihood. In addition to wood and food, the land grows families, democracy, passion, biodiversity, wildness, safety, and art. He said we must "cease being intimidated by the argument that a right action is impossible because it does not yield maximum profits, or that a wrong action is to be condoned because it pays. That philosophy is dead in human relations, and its funeral in land-relations is overdue."[4] Leopold's science, poetry, and practice are full of efforts to harmonize the material and cultural harvests with the integrity of ecological systems.

Leopold's emphasis on community is important.[5] Communities unify members: membership creates obligations as well as providing the benefits of protection, opportunity, and identity, and membership requires respecting and protecting qualities that define the community. Members must limit behaviors that damage the community. Membership also requires good-faith participation in negotiations that decide the norms of behavior as well as the privileges and responsibilities of community membership. Members need to explicitly negotiate their visions for a thriving community: What defines a good life? What counts as acceptable environmental quality? What ideals should be passed on to future generations? Which social institutions and behaviors will create and sustain these desired conditions? Who or what deserves membership? What rights, privileges, and responsibilities do members deserve? These negotiations unify community members by shaping shared expectations.

Fundamentalism narrows the decision space where common interests can be found by promoting a politics of blame and shame when what is needed is deliberation and collaboration. Worse, it encourages allegiance rather than understanding. Fundamentalism discourages appreciation and respect for the many natures that evoke hope, wonder, and political action. This book pluralizes nature in hopes of moving us beyond environmental fundamentalism toward the critical tasks of understanding and sustaining the many natures that add immeasurably to the identities and qualities of our lives.

It is common in environmental debates for "nature" to be used in defense of policies and positions. It is rhetorically effective to justify a particular outcome because it is "natural," has "natural causes," or was "naturally selected" or because "nature knows best." But these appeals invoke the naturalistic fallacy: just because something occurs in nature does not mean that it should.[6] Some sociobiologists, for example, argue that evolution predisposes human males to be promiscuous because such behavior supposedly increases the number of descendants and therefore the "success" of a promiscuous man's genes; this contention, if true, does not mean that modern culture should sanction infidelity, adultery, and rape. Some evolutionary psychologists argue that evolution has favored humans who were aggressive in their pursuit of food, shelter, and mates; this contention, if true, does not mean that modern culture should promote murder, warfare, and confrontation as the primary means of dispute resolution. Likewise, some social Darwinists suggest that competition for survival improves society's fitness; this contention, if true, does not mean that modern culture should discourage welfare, public education, and other social practices that increase opportunities for the socioeconomically disadvantaged. Other invocations of nature's moral authority are used in arguments for and against abortion, homosexuality, and the subjugation of women. In most cases, the characterizations of nature are partial and biased; in all cases, they are insufficient. The values and qualities we choose to define and guide our communities must be openly deliberated, not ignorantly justified by trite appeals to nature. Recognizing many natures mitigates against the naïveté and absolutism that hinder open deliberations.

"Nature" is also used to describe desired environmental conditions and goals of environmental management: "keep it natural," "mimic nature," or "don't exceed the natural range of variability." Scientized synonyms of nature can include "biodiversity," "ecological integrity," and "health." These references to a singular nature are ambiguous and misleading. Pluralizing nature forces us to be more specific about the conditions we desire and why they are desired.

Environmental fundamentalism collides with other worldviews such as religious fundamentalism to produce sparks that fuel the flames of polarization and paralysis. A popular environmental critique of Judeo-Christian traditions argues that teachings of dominionism and dualism embedded in the Bible are responsible for the tendency toward environmental exploitation found in Christian Western civilizations. Alternative interpretations of Judeo-Christian traditions emphasize and mobilize concern about caring for God's creation. The pluralizing of nature that occurs in subsequent chapters reveals enormous overlap between religious and environmental causes. Goals of environmentalism and organized religion likely have more commonalities than differences: they share common concerns about the environment, the economy, culture, and the future. The same can be said for ecologism, primitivism, capitalism, and other absolutisms.

PLURALIZING NATURE

Pluralizing nature facilitates the collaboration and deliberation that resolves environmental conflict and implements solutions. If stakeholders in community-planning efforts believe that their nature is the only possible nature, then they may fail to recognize interests they share with others and instead waste scarce civic energies defending the legitimacy of their nature. The better we understand the many natures that can exist, the better we will be able to comprehend the concerns of others, evaluate their positions, and find common ground. Resolving difficult environmental conflicts requires collaborative, deliberative efforts where participants comprehend and respect one another's perspectives, and when conflict seems intractable, are able to reframe the conflict around issues and interests stakeholders can agree on.[7] Pluralizing nature increases the decision space where acceptable conditions may be found as well as the political support for these conditions by including more community members who care and giving them more reasons why they should care. The increased number, intensity, and overlap of interests might help mobilize the political support needed to create thriving and sustainable communities.

As an example, take General Grant, a sequoia tree located in Kings Canyon National Park, California. The third largest tree in the world, it towers more than thirty-six stories in height and has a diameter exceeding forty feet. Several thousand years old, General Grant is the largest living memorial to Americans who died in war. Its existence likely can be traced back to Native American burning practices because sequoias require fire to release seeds, prepare the soil, and open the forest canopy. General Grant reminds people of their ties to these Native Americans and to the early pioneers who settled the continent. And it has value to the Judeo-Christian God, who created trees and other vegetation on the third day and declared them "good."

It could provide decent jobs to hardworking people who pay local taxes that in turn build schools and hospitals if it were felled and sawed into boards to build houses. It might even host genes or chemicals to cure debilitating diseases. It does retain water, sequester carbon, purify air, and perform a myriad of other ecological services on which the lives of countless insects, bacteria, and other forms of life depend. It evokes awe and humility in most people who see it and is photographed and commemorated in countless coffee-table books and family albums. People stroll beneath its branches and ponder deep thoughts about life, ecology, and their place in the universe. They search for lessons about how to live a worthy life. The tree's magnitude inspires still others to pursue careers as foresters or conservation biologists. Moreover, it is alive. It grows, seeks resources, and resists infections. By these actions, the sequoia demonstrates a will to live. Perhaps this willfulness to survive has value in and of itself, independent of any human value. There are indeed many ways to know and appreciate General Grant. The more we broaden and deepen our interests in this tree, and in nature generally, the more we will be motivated to engage one another in efforts to create and sustain communities where these interests thrive.

Pluralizing natures mitigates the polarization and paralysis of environmental fundamentalism because it asks us to respect and critically examine diverse interests. I am not advocating what has been called a "flabby" pluralism that uncritically accepts all perspectives, paradigms, and vocabularies. Nor am I advocating what has been called a "fortress-like" pluralism where isolated groups work to build and advance their internally consistent methods and theories, unwilling to communicate across paradigm boundaries and unable to agree on universal standards of truth. Rather, I'm suggesting an "engaged" pluralism that accepts multiple perspectives on every issue and replaces the quest for certainty and absolutes with the negotiation of truth and objectivity through agreement.[8]

Engaged pluralism demands respect and responsiveness to the positions and understandings of others. It requires active listening and a critical and self-reflective thinking that acknowledges the uncertainty and limits of knowledge. Applied to environmental negotiations, engaged pluralism seeks ways to understand, value, and illustrate the environmental conditions that describe where we want to live in the future as well as mobilizing the political support needed to get us there. Engaged pluralism is characteristic of a healthy democratic society where we deliberate our ideals and standards of conduct.

U.S. citizens insist on freedom of speech, private property rights, democracy, and related ideals codified in the Constitution and the Bill of Rights. Ideals such as these have been negotiated over centuries and enforced by wars. We violate our ideals only after serious political debate, no matter how efficient or profitable the violations may be. These ideals create our legacy; we hope and expect that our descendants will know and appreciate them. Among these ideals are the environmental qualities that define us. The claim that America is "nature's nation" is well founded.[9] American culture has been shaped by the American environment. The environmental qualities we choose to create will decide not just the sustainability of our communities, but also our identities.

The primary hope of this book is that pluralizing nature will help us identify, care about, and sustain a thriving, unifying Leopoldian community. A secondary purpose of this book is to dispel certain myths and misperceptions that pervade public negotiations about environmental quality. My intent, in this regard, is to cut out some of the flabby pluralism and strengthen the descriptions and justifications of our desired future conditions so that they appeal to an ever-widening circle of stakeholders. Public discourse about our future is weakened and cheapened by myths about a balanced nature, a pristine continent, a noble savage, and a purposeful evolution.

So-called environmentalists invoking these holy grails paralyze discussions, waste valuable political capital, and deflect needed public discussion away from widely shared public goals of thriving and sustainable communities. The so-called economic exploiters find fresh meat for their dismissal of environmentalism and understandably look elsewhere for definitions of environmental quality and ideals about future communities. Discounting and dismissing the whole environmental message becomes easier if a few prominent rhetorical points are so easily questioned. Bestsellers such Michael Crichton's *State of Fear* effectively throw out environmental babies because of legitimate criticisms about dirt in environmentalism's bathwater. *Infinite Nature* advocates pluralizing nature in order to increase the decision space where acceptable environmental conditions may be found, but I do not uncritically accept all natures and intentionally single out some construals to dismiss as myth and misperception.

Some worry that pluralizing nature diminishes the legitimacy of environmental protection and hastens environmental degradation. The relativist argument goes something like this: Because there are many natures, all natures must be equally acceptable. I disagree. That type of thinking is flabby pluralism. Engaged pluralizing encourages responsibility. It forces us to accept our role in determining which natures exist. Pluralizing nature forces us to be explicit and deliberative about our choices among the many natures that have been and might be. Pluralizing nature does not suspend biological realities; rather, it erodes the myth that our knowledge of nature exists independently of our agendas.[10]

ABOUT THE BOOK

Our understandings of the world are socially constructed. What anyone knows about nature depends as much on human theories, language, and cultural context as it depends upon what is "out there" to be felt, tasted, viewed, and heard. The social constructivist assumption, as it is applied here, does not question whether reality exists independent of human thought or whether science can describe "facts" that help us understand that reality. For example, I do not doubt that trees exist or that they grow, die, and fall independently of humans. How the fallen tree is described, however, does depend on humans. We could describe the sound of the falling tree using decibels. We could just as easily describe the increased sunlight now reaching the forest floor, the decreased habitat for canopy-dependent insects, the increased habitat for decomposition fungi, the lost economic value of timber, the increased access to a scenic vista, or the educational opportunity of witnessing ecological disturbances. Any one of these descriptions, and many others, accurately characterizes facts about the fallen tree. How we interpret the fallen tree is shaped by the measurements and definitions we apply to it. We are limited in how we can know the fallen tree, or any other aspect of nature, by the methods and purposes we have to describe it. In this way, nature is socially constructed.

I also believe it is impossible to separate descriptive from evaluative aspects of any discussion of nature, scientific or otherwise. The only reason I am able to discuss the natures in this book is because people have cared enough to create a vocabulary conducive to their description, understanding, and appreciation. Nature is infinitely complex and we ask questions about only a small portion of the infinitude. We invent terms and conceptual models to serve as abstractions to represent the parts that interest us, the parts we value. We know little or nothing about parts of nature's infinitude that fall outside these realms.

In addition to crossing the fact-value divide, I deliberately wade across disciplinary boundaries. Disciplines deepen our understanding of particular aspects of nature, but they necessarily ignore other aspects. Disciplines are penetrating spotlights that focus and magnify our gazes so that we may see further and finer. By necessity, a single spotlight casts a narrow beam and much falls in its shadows. I have tried to use multiple spotlights, and any other source of insight I could muster, to map the broad terrain of many natures. In some ways, I am now making excuses and asking the experts' indulgence for glossing over important nuances and understandings available only through prolonged disciplinary analysis. While pluralizing nature is somewhat novel, the content of this book is not. It is drawn from established sources in environmental history, ethics, and philosophy, as well as various natural sciences. I've tried to provide endnotes that direct interested readers into the detailed and voluminous literature associated with each nature. In this work, I merely enter the forest and must leave further exploration to you.

Some readers will be frustrated that I do not advocate *the* natures that should be sustained. Even within a chapter specific to a nature, I intentionally discuss multiple, sometimes conflicting ways of knowing and valuing that nature. Only the affected community can decide through study and deliberation which of these many natures should be created and sustained. This does not mean that I uncritically accept all natures as equally deserving of consideration. The reader will find me critical of many interpretations that I think have been refuted as myths or uncovered as disguises for political power. The purpose of this book is not to advocate a "correct" nature. Rather, the purpose is to partially map the terrain of possible natures, to legitimize multiple natures and multiple values, and to encourage that these natures and values be welcomed and considered during deliberations that shape our future.

Anthropogenic Nature

Native Americans · Noble Savage Myth
Modern Humanized Biosphere

NATIVE AMERICAN LAND MANAGEMENT

The Spanish conquistador Hernando de Soto went to the New World determined to find fortune and fame several decades after Columbus discovered the Americas. Other explorers preceded him, but de Soto's was among the most ambitious and best documented of the early expeditions. His seven-ship armada set sail from Cuba and arrived in Florida near Tampa in 1539. He commanded an army of over six hundred men provisioned with hundreds of livestock. They wintered in Tennessee, and some reports have them marching as far north as Chicago in search of a northern sea that would provide access to China. De Soto "discovered" the Mississippi River and traveled as far west as Texas before he died. In their four years of travels, the army encountered a cultured population and a domesticated landscape, but not the gold, gems, and other riches hoped for by the Spanish crown. Traveling along well-worn paths that connected numerous villages and several large cities, they commandeered provisions from thousands of acres of irrigated fields sprouting corn, squash, and other domesticated crops. Willing and unwilling natives served as guides and laborers.[1]

Estimates are difficult and dangerous, but some suggest between 4 and 7 million people lived in North America at that time, and perhaps 10 percent of them lived in the area de Soto explored (estimates go as high as 100 million if we add Central and South America).[2] A small and scattered population by today's standards, but certainly the continent was neither empty nor pristine. European horses, armor, metal, and military tactics overwhelmed Native American defenses, but more importantly smallpox, influenza, measles, typhoid, mumps, and other European diseases devastated the Native American population within a generation or two. More than 90 percent of Native Americans perished in the century following Columbus. They lacked biological resistance, and they lacked the knowledge, learned in Europe through harsh lessons taught by plagues, that surviving outbreaks of infectious disease requires quarantine of infected communities. Most of Native American culture, economy, language, technology, and agriculture perished as well.

The human population decline left huge voids in the North American ecology and created an appearance for later European explorers of a pristine, abundant, and empty wilderness ripe for the taking. Absent a thriving human culture, Europeans had an easier time claiming land for God and crown. European-based American culture has long preferred a cultural narrative that begins with the myth of discovering and taming pristine wilderness populated by a few heathen savages. But the myth is easily dispelled. The existence of sophisticated cultures and extensive land management prior to European discovery should be incorporated into the stories we tell ourselves about our cultural origins and into our land-management ideals. Several additional illustrations of the cultivated pre-Columbian American landscape should suffice to make this point.[3]

The Native American city of Cahokia prospered east of current-day St. Louis. Sometime around 1000 AD, when Cahokia civilization was at its peak, the city was larger in size, population, and complexity than London. Enormous earthen mounds strategically located along the flat terrain designated social status, religious symbols, and economic trade locations. The largest, Monk Mound, contained some 22 million cubic feet of soil, rose almost one hundred feet, covered more than fourteen acres, and rivaled the total volume of the largest pyramid in Egypt. It was topped with a massive ceremonial building, perhaps the location of religious and state services and residence to the high chief. Well over a hundred other mounds could be seen from Monk Mound, many of which remain preserved. These mounds defined the central city of this empire in which twenty to thirty thousand people lived behind a protective two-mile-long timber wall.

Cahokia was connected by trails to smaller cities, suburbs, and remote agricultural villages located up and down the Mississippi, all sharing a common religion, culture, and economy. The trade network extended even farther, and it brought copper from the Upper Great Lakes, mica from the southern Appalachians, and seashells from the Gulf of Mexico. Technological advancements included a solar calendar, fishing nets, spears, stone-cutting and grinding tools, and, importantly, three kinds of hoes that made food production much more efficient. Agriculture based on domesticated corn, squash, sunflowers, and pumpkins planted in the fertile Mississippi floodplain provided a reliable and abundant food supply to enable and sustain a culture that produced numerous artifacts, art, leisure activities, and a specialized workforce. Cahokia had all the marks of a great civilization: music, dance, sport, worship, complex social organization, city planning, advancing technology, long-distance trade, abundant and controlled food sources, and a large population.[4]

The economic and cultural accomplishments of Cahokia are impressive, but they pale in comparison to the Aztec empire, located in modern-day Mexico. By 1500 AD, at about the time Columbus was sailing the ocean blue, the Aztecs had allied with or conquered other native cultures and seemingly controlled over 100,000 square miles of territory and as many as 10 million people. Tenochtitlán, the capital of this state, was built for defensive purposes on an island in a lake. It housed 200,000 people, making it much larger than either London or Paris and one of the most heavily populated places on Earth at the time. The Europeans who "discovered" it were awed by its size and beauty and lusted after its wealth. They found the city's large lake crowded with boats engaged in pleasure and business. The central district, with its brightly colored houses and large palaces, supported markets, art, commerce, and a zoo that exceeded any known in Europe. Strategically located islands had been built from rich soil dredged from lake bottoms. Causeways connected the island city to the mainland, and a grid of canals and roads organized city transportation. An aqueduct carried fresh water into the city, and a thousand-person sanitation crew collected garbage and human waste and transported it out of the city to be processed and recycled as agricultural fertilizer.

The land surrounding Tenochtitlán evidenced sophisticated agriculture. Terrace fields extended into the lake's edge to capture water and soil fertility. Swampy land had been drained or raised, and hillsides were terraced for water management and erosion control. A system of dams, canals, and aqueducts provided irrigation. Even to the well-traveled Europeans—some of whom had been to Rome, Constantinople, and Venice—this North American city was unprecedented in its complexity and beauty. In the Europeans' favor, however, were distinct advantages in social organization, writing, weaponry, and, most importantly, germs.[5]

Europeans discovered in North America a nature that was cultured, not pristine. Most explorers of this supposed New World rarely blazed new paths. They followed existing roads, provisioned their parties at established villages, and profited through trade and interaction with thriving economies. Native Americans had already altered the environment with heavy hunting, fire, agricultural turning of soil, and importation and domestication of species. Mast nuts and fruit-producing trees were planted, pruned, thinned, or otherwise encouraged. Intentional fires cleared vast swaths of forest, creating habitat and hunting grounds for deer, elk, and buffalo. Cities, irrigation, roads, and cultivation dramatically transformed large portions of the continent.

Not all Native Americans lived in dense, hierarchical, agricultural communities. Many were hunter-gatherers, but their population still applied vast pressure on edible flora and fauna. Millions of people over thousands of years changed the evolutionary and ecological trajectories of North America. Some of these impacts were obvious to European settlers, such as waterworks, cities, agricultural fields, and religious symbols. Other impacts were less obvious; there was no way to know that fields and prairies had been created by years of regulated fire, that large mammals had been hunted to extinction, that some plants were scarce because of intense gathering pressures, and that others were abundant because they had been planted and nurtured.

The North American nature that we know today is the result of many generations of human management, most of which predate European settlement. Yet the nature that European pioneers discovered and later romanticized was largely unpeopled. Europeans found a continent fully stocked and ripe from the labors of generations of Native Americans, but without the significant native population with which to compete for moral claims of ownership or with which to share the bounty. Unknown to most European settlers, their wilderness paradise had been won for them with what today would be called an act of biological warfare.

WERE NOBLE SAVAGES LIVING IN HARMONY WITH NATURE?

The environmental conditions of pre-Columbian America are sometimes used in environmental science and policy as benchmarks of healthy, functioning ecological systems and as goals for ecological restoration and wilderness management. Idealizing these conditions presumes that modern Americans have a lot to learn from Native Americans, their land ethics, and their land practices. It presumes that Native Americans lived in harmony with nature and therefore possessed a sustainable land ethic that motivated sustainable land-use practices. These presumptions may be misplaced.

Do aboriginal cultures learn through trial and error to live in harmony with the land, to respect Mother Earth, and to live within the limits of local nature? Are humans, in our natural state, selfless, peaceful, and untroubled? The inflammatory term "noble savage" reflects tensions embedded in this debate. Modern cultures have a love-hate relationship with "primitive" cultures that has ebbed and flowed through time. At various points in our history, aboriginals were thought to be subhuman savages. Such reasoning removed moral obligations normally granted to humans, allowing Europeans to enslave and subjugate "primitive" people without guilt.

"Primitive" people, as the adjective implies, were considered less evolved and not governed by the same biological and cultural rules.

With the passage of time and the advance of "progress," this negative opinion of aboriginal people would change. As Western cultures powered by a capitalist economy and inventive technology marched across the continent, they left behind some unwanted consequences, such as barren, inhospitable, smoke-filled wastelands. In addition, many people feared that capitalism and modernization were damaging the human spirit—turning us into selfish, competitive, pleasure-seeking drones. As the negative side effects of industrialization, urbanization, and modernization came into focus, some people began to look to aboriginal cultures for possible solutions. The savages came to be seen as "noble."

So-called Romantics looked to "primitive" cultures for sustainable and meaningful lifestyles. Literature about Mother Nature, nature's children, and primitivism found a receptive audience. Boy Scouts and children's camps taught Indian lore and values. George Bird Grinnell, an important turn-of-the-century conservationist and advocate for national parks, published an article in 1916 in *Forest and Stream* entitled "What We May Learn from the Indian." In it he wrote that the Native American "protected the game on which he depended and practiced methods of economy in hunting that American sportsman may well take to heart." Modern land managers, and society generally, supposedly have important lessons to learn from Native American land-management practice and philosophy.[6]

Native American stories glorifying nature and portraying a harmonious, balanced relationship between culture and nature are ubiquitous. But at least some of these portrayals better reflect the romantic storytelling needs of Hollywood and Disney than they do reality. For example, a frequently cited quote attributed to Chief Seattle appears in such noted publications as former vice president Al Gore's *Earth in the Balance* but is really a quote from a long-forgotten Hollywood script.

This we know—the earth does not belong to man, man belongs to the earth. All things are connected like the blood which unites one family. Whatever befalls the earth befalls the sons of the earth. Man did not weave the web of life; he is merely a strand in it. Whatever he does to the web, he does to himself.[7]

Claims that aboriginal peoples left small ecological footprints seem easily muted, at least as a general rule. De Soto's discoveries and the Cahokia and Aztec civilizations leave little doubt that the ecology of the Americas was the ecology of Native Americans. While there might be instances of indigenous people living in harmony with nature, there exist many examples of exploitative and unsustainable behavior. Throughout North America there is evidence that herds of deer, elk, and buffalo were driven off cliffs, killing far more animals than could be eaten or used. Excavations of these sites show that only the top few animals were butchered for food and that animals at the bottom of the piles were left to rot. Large mammals such as camels, woolly mammoths, mastodons, and giant ground sloths went extinct soon after humans populated North America. Climate change is partially to blame, but so are hunting pressures by the very first Native Americans. In addition, native farming practices cleared huge swaths of land, eroded soil, depleted nutrients, and increased salination. Environmental histories show that wood supplies—critical for building, heating, and cooking—were exhausted in many regions, likely contributing to the decline of civilizations.

Analyses of animal bones and shells found near archaeological sites where aboriginal people once lived show a gradual decline in the size of most hunted animals, followed by their increasing scarcity, and eventually local extinction—presumably due to unsustainable hunting-and-gathering practices. Analyses comparing human skeletal remains over several generations in the same locations show increasing malnutrition and starvation due to decreasing food sources. Rather than living in harmony with Mother Nature, archaeological evidence suggests instead that many aboriginal cultures were exploitative and opportunistic.[8]

Archaeological records document thousands of "primitive" cultures that for brief periods managed a semblance of balance with their environment but, sooner or later, overhunted animals, degraded their surroundings, and undermined their own existence. Whole villages would regularly relocate, seemingly because of resource degradation. In areas with low human populations, this migratory lifestyle was sustainable. The degraded areas would recover after several decades or centuries, allowing humans to return. In areas with higher population densities, such migrations would cause violent territorial disputes and/or cultural collapse. The famous Anasazi civilization, builders of multistory villages in Chaco Canyon and dramatic cliff-side dwellings at Mesa Verde, supposedly declined and vanished because of changing ecological conditions, some of which might have been climate related, but much of which was self-inflicted. A similar argument is made about the Cahokia civilization discussed above. These cultures dissolved in part because they had depleted their forests and degraded their land's productivity.[9]

A dramatic example of ecological collapse can be found at Easter Island, the easternmost island of Polynesia and perhaps the last of these islands inhabited by humans, who reached its shores around 300 to 400 AD. Easter Island is triangular in shape and approximately sixty-four square miles in size. Settlers from other Polynesian islands brought with them wood- and stone-carving technologies, social and economic organizations, written script, as well as domesticated plants and animals such as taro, yam, sweet potato, banana, sugarcane, chickens, and rats. At the time of human settlement, the island was covered with palm trees and lush vegetation and was perhaps Polynesia's most productive breeding ground for seabirds.

Forests and fields were cleared and planted with domesticated crops during the first several hundred years of human habitation. The island's rich natural resources included edible plants, fertile soils, and ample materials for building a sophisticated civilization that included leisure, craft, and ritual. Trees were fashioned into canoes for deep-sea fishing, and their wood was used to build homes, make art, cook food, and transport and erect the island's still famous statues. At the civilization's peak, around 1300 to 1500 AD, approximately eight thousand people lived on the island. Enough free time, food calories, and social organization existed to allow construction of more than eight hundred giant stone statues. The meaning of these statues remains elusive, but they clearly exhibit a level of craftsmanship and social organization that remains impressive to this day. Weighing as much as eighty-five tons and hewn from solid rock, the statues were painstakingly transported considerable distances from their inland quarries to the island's periphery and other sacred spots and then raised into upright positions. The trees, no doubt, played a critical role as rollers and levers in such activities. Some estimates suggest it took fifty to seventy-five strong, well-fed men five to seven days to move and erect a typical statue, wrapped in rope and secured in a wood frame, to its final ceremonial location.

By the mid-1500s, Easter Island became deforested and soil fertility declined. Crop yields dropped dramatically. Freshwater lakes became contaminated and filled with sediment, perhaps from quarry operations. The native bird population collapsed as they and their eggs proved too accessible to humans and the rats that traveled with them. Deforestation produced a shortage of wood for canoes, which in turn caused fishing yields to plummet and reduced protein in the typical diet. Social organization deteriorated, and statue construction ceased. Warfare, cannibalism, and slavery became prevalent, and available resources were channeled to weapons manufacturing. People moved from their communities and houses located within easy access to water and food into remote but more easily defended caves. The grand statues were toppled and decapitated, perhaps by warring rival factions, and human population declined.[10]

It seems this thriving and intricate civilization collapsed, at least in part, because of environmental mismanagement. Even with the ocean's bounty, humans could not sustain themselves on a fixed land resource. Rather than living in harmony with the land, people exhausted or polluted the environmental qualities on which their lives depended. Hindsight suggests the culture's self-interest would have been better served by protecting key environmental capacities such as the forest, soil, bird breeding grounds, and fresh water; but the people were unable to see or act on the relationship between their self-interest and environmental quality. Other environmental histories tell similar ecological horror stories after human settlement and raise serious questions about the myth of ecologically noble savages.

HUMANIZED BIOSPHERE

Aboriginal cultures generally, and Native American cultures specifically, might not have lived sustainably, but they placed fewer demands on nature than do modern capitalist cultures, which, assisted by better technology, impose much more rapid environmental changes over much larger areas. The American landscape changed quickly and dramatically with application of European technology such as cows, plows, horses, and harnesses. Imported from Europe, these technologies dramatically increased the amount of earth that could be turned, the number of trees hauled, and the amount of resources extracted. Harnessing fossil fuels and combining them with mechanical apparatus further magnified the amount of change one person could create. What once took decades of human labor, and then years of animal labor, could be accomplished in days by one ingenious and motivated person with bulldozers and chainsaws.[11]

By some estimates, 40 to 60 percent of the biosphere is directly manipulated to satisfy human needs. The ecological implications of a humanized biosphere are many and varied, so several illustrations should suffice. Humans use over half of the accessible fresh water on Earth and have transformed nearly half of the land surface through agriculture, forestry, and urbanization. Approximately one-fourth of bird species present at the dawn of humanity are now extinct, and favored species such as corn, wheat, rice, cattle, bees, and tulips have been domesticated, transported, planted, and nurtured so that they grow and live just about everywhere on Earth. Other species have stowed away in the ballast tanks of ships, in untreated wood products, in food, and on the leaves and in the soil of transplanted plants. The most hardy and invasive of these introduced species outcompete native species and have started a biological revolution that will transform American ecological systems in perpetuity. Some of the consequences are devastating. The Chestnut blight, for example, reconfigured the entire eastern forest—effectively removing the dominant species.

The list of our impacts is impressive and extensive.[12] We are an industrious, ingenious, and invasive species. But by no means has nature been dominated by human ingenuity and technology. Tsunamis, earthquakes, droughts, hurricanes, and genetics are just a few examples of the power and dynamics of nature that remain well beyond human understanding and control. It is difficult, however, to think of one thing on Earth that has not been altered by humans. Go ahead and try. Even plate tectonics, ocean tides, and the moon's orbit might have been imperceptibly altered by damming large bodies of water in the Northern Hemisphere. These dams have changed the distribution of water (and mass) around the circumference of a rotating object and thus necessarily, but negligibly, altered Earth's rotation. Even the deep, dark recesses of the ocean floor have most likely witnessed chemicals diffusing from ocean dumping of spent nuclear fuel, toxic wastes, and household sewage.

CONCLUSION

Native and modern American actions have so changed the landscape that some observers lament that nature has disappeared—transformed by human enterprise.[13] Clearly, nature has not literally vanished or died. Ecological processes continue to sustain life, and evolution continues to create new life. But humans have domesticated so much of Earth that nature can no longer be understood as something independent and autonomous. It is probably impossible to delineate where humans stop and nature begins. Is this a problem?

Land-use goals that seek authentic, pristine, and pre-Columbian conditions become problematic in the context of anthropogenic nature. Humans and nature have co-evolved for thousands of years, the two histories fully intertwined. We don't have the means, even if we had the will, to restore and maintain conditions at prehuman conditions. Human-induced changes are now part of nature, and we must employ land-use goals that acknowledge this fact. Accordingly, evolutionary and ecological science are moving beyond static ideas of an authentic and pristine nature, developing concepts and theories that explain a dynamic, hierarchical, and humanized nature. Concepts such as resilience, sustainability, and ecological health offer possible alternatives to pristine and wild as land-use goals. These new understandings offer alternative means to frame concerns about land-use management without involving pre-Columbian conditions as benchmarks.

Many people still value authentic, prehuman conditions, in spite of or perhaps because of mounting evidence of a humanized biosphere. These conditions might be valued for spiritual reasons (chapter 9). If we view nature as God's creation and as a conductor of divinity, we might worry about soiling the virgin and desecrating the altar so that we avoid a second eviction from Eden. Alternatively, some people might value the rights of nonhuman plants and animals to live an untrammeled life, free from human control and interference (chapter 11). Of course, others might value dehumanized nature simply for aesthetic reasons (chapter 12). Vast tracts of primeval nature evoke awe and wonder. Nature is different from human-made art in that hiking and fishing are experienced through immersion rather than observation. People also find meaning and identity in nature (chapter 13). They believe nature transcends human culture and provides access to deeper truths. Nature is a mirror into which humanity looks to see itself and its history and to assess its future. Admitting humans have changed and perhaps created nature removes our external basis of comparison: it cuts off our anchor and sets us adrift in the ocean of relativity as we search for truth and purpose. These and other understandings of nature provide reasons for valuing so-called authentic, original, and wild nature. They are the topics of subsequent chapters.

Should modern humans be ashamed of our interactions with nature? Have we fallen from grace and lost the ability to live in harmony with nature? Our history gives ample reason for concern. Humans have rarely acted sustainably, and unsustainable behavior may be at its peak today. The power and reach of our technology allows environmental change of unprecedented scope and scale. Yet unlike previous cultures, we are aware of the extent of environmental change and we openly debate factors affecting sustainability. There is good reason to be hopeful. Rather than lament that modernity has lost insights once possessed by "primitive" cultures, we should feel proud that we now recognize what so many previous cultures ignored, at their peril. Perhaps if we bring positive sentiment rather than guilt to our deliberations, we will proceed with greater haste and with a greater sense of purpose. Rather than feeling bad about what we have lost, we should feel good about what we can accomplish. And there is much to do.

Evolving Nature

Dynamic · Naturally Selected
Random · Lucky Humans · Biodiversity
Taxonomy · Social Darwinism

Darwin's theory of evolution revolutionized our understanding of nature. Life on Earth is old—very old by human standards, perhaps 4 billion years old—and it is constantly changing. Life, Darwin argued, changes through random and unpredictable events; it is not the perfected creation of an all-knowing, all-caring God. Darwin demoted humans from the top rung of the evolutionary ladder and showed that all life on Earth shares a common origin, traceable through long trajectories of incremental change—we are all members of the same family of life. Rather than being divinely predestined to emerge from the evolutionary process, humans, like other species, are fortuitous, unrepeatable accidents of history.

Not everyone believes in evolution or in its serendipitous qualities. Genesis offers an alternative explanation of creation, one we will examine in chapter 9. Nonetheless, nature known through the lens of evolution holds a privileged place in the environmental debate. It provides a creation story and structures the way many people comprehend and speak about nature. It is the foundation of most biological sciences, so what is known about nature through the lens of evolution fills volumes. This chapter focuses on just two of the many possible evolutionary understandings of nature: (1) a nature that is dynamic and (2) a nature that is random, not progressive. The chapter concludes by questioning several environmental and social policies justified using inferences from evolution.

DYNAMIC NATURE

The dramatic and continuous changes that characterize life on Earth stand out in bold relief when nature is viewed from the perspective of geological time.[1] Species migrate, evolve, or go extinct in response to climate change, plate tectonics, and competition. Lakes and inland seas rise and disappear. Rivers appear, erode valleys, change course, and disappear. Soil is formed, eroded, buried underwater, and raised again perhaps as soil or compressed into rock as parent material for future soils.

The atmosphere's composition has changed dramatically. Methane and ammonia dominated Earth's early atmosphere, while the oxygen on which so much current life depends was virtually nonexistent until early life-forms perfected photosynthesis. It was probably not until 600 million years ago that the concentration of aerobic life-sustaining oxygen and DNA-protecting ozone reached concentrations similar to those found today. This new atmosphere moved evolution along paths to the biota we know today.

Looking back through the lens of geological time, we see many natures, not just one nature. The nature present in the area we call Chicago, for example, existed as mowed city parks, Native American burned savannas, ice age glaciers, stomping grounds for Jurassic dinosaurs, the submerged floor of a sea separating the east and west coasts of what is now North America, a host for abiotic bacteria struggling into existence on the supercontinent Gondwanaland once located well south of the equator, and, for the first billion years or so of Earth's history, a molten, asteroid-battered rock void of life.

Which of these previous natural states is better than another? This is not a trivial question. The Chicago area has an active and effective ecological restoration effort that seeks to replace thick local forests dominated by exotic (human-introduced) species with the oak-savanna prairies that existed prior to European settlement.[2] Chopping, burning, and replacing the "wrong" vegetation with the "right" vegetation has generated considerable controversy. Many Chicagoans, who consider themselves environmentalists and nature lovers, object to the removal of a forest that now fills their views and shades their walks. What justifies killing helpless trees and the animals that depend on them? Why is the "native" savanna better than the "exotic" forests? The restorationists are bewildered that their motives should be doubted. To them it is beyond question that natives are better than exotics and that the nature existing prior to European settlement was in most every way healthier and superior to the unnatural, exotic-filled, fire-suppressed, managed nature that now dominates the Chicago landscape.

So we return to the question: Which of the many previous natures is better than another? Which nature should Chicago's active ecological restoration efforts strive to restore? Evolutionary science does not provide a clear answer. It seems obvious that human inhabitants of Chicago would want one of the natures existing during the last 5 million years (or so) when humans diverged from chimpanzees, gorillas, and orangutans. For that matter, they probably would prefer to wait until *Homo sapiens* emerged as a distinct, upright, tool-using, language-capable species a million or so years ago, or even until after "modern *sapiens*" evolved within the last 200,000 or so years. Also, given that many of Chicago's residents would rather not abandon their homes and business to glacial moraines, they probably do not wish to live in the natural conditions that existed prior to the retreat of the most recent glaciers, which did not occur until roughly 15,000 to 8,000 years ago.

The postglacial time period still leaves many natures to choose from. Dramatic changes occurred across North America during the last few thousand years. Huge ice dams prevented water in the Great Lakes from flowing down the St. Lawrence Valley and into the Atlantic. Instead, it flowed south, down the Mississippi River, and increased the flow of that river to four times that of any flooding in recent history. When glaciers retreated enough so that the flow down the St. Lawrence could resume, icy waters rushed into the Atlantic, disrupting warm ocean currents and changing worldwide weather patterns.

As glaciers retreated, so did their moderating effects on summer temperatures. Glacial mountains were no longer miles high, so they no longer redirected prevailing winds. Windstorms became more destructive and summer heat more extreme. Weather patterns became more varied; years of drought oscillated with rain and flooding. Water, previously locked in ice, fueled the cycle of rain. So much water returned to the oceans that sea levels rose 300 to 500 feet, reclaimed 30 to 60 miles of costal plains, and drowned vegetation that had migrated there. The sheer weight of the ice pack had depressed the land, which has been rebounding ever since. Hudson Bay, for example, is still rising at about half an inch per year and may drain when its bottom eventually elevates above sea level.

Vegetation followed the retreating ice. It was once common to think of vegetative communities responding cohesively to forces of change, but now it is more accepted that species respond independently to environmental conditions, and that the systems of species we find in any one spot are really just collections of individuals that happen to be passing through. Species migrate at different rates, each with different tolerances for soil, temperature, moisture, and other species. Thus "natural" ecological communities we find today around Chicago and elsewhere did not exist in the distant past and likely will not exist in the distant future.

Trees, rooted in soil but dominating human perception of the landscape, provide a case in point. Hemlock, poplar, birch, and maple launch their seeds into the winds that transport them great distances. Oak, beech, hickory, pine, and chestnut migrations are powered by squirrels, birds, and gravity. All tree species fled south as glaciers scoured Chicago. They had taken refuge in the Appalachian foothills and on the coastal plains of Virginia and North Carolina that are now under seawater. As trees go, the white spruce is aggressive and nimble. It closely followed the retreating glaciers. White pine and eastern hemlock followed next. They expanded north and west, moving toward Chicago more slowly than spruce, but have now caught up with it to enjoy approximately the same northern range. Oak had retreated much farther south and west than either spruce or pine, but eventually caught up and beat the pine's migration by a thousand years into its northernmost range. The hemlock, migrating slower and along a different path, took another thousand years or so to catch up. The prairies so valued by Chicago restorationists, meanwhile, had retreated all the way to Texas before returning to create the Great Plains.

Dramatic changes occur most everywhere in response to changing climate. The southwestern United States, for example, was moister and cooler than it is now. It was dense with conifers that retreated from higher, windier, and colder ice age conditions. The low-lying deserts of the Southwest, now among the hottest and driest places on Earth, filled with massive freshwater lakes from glacial melt. They gradually evaporated over thousands of years, leaving the salt flats we find today.

Massive dams of glacial moraine and ice created temporary lakes that drained abruptly as the ice melted and dams burst. Pulses of water, hundreds of feet high, reworked river channels. Few things would have been more dramatic than the explosions of water coursing down the Columbia River system. Glacial melt collected to form a 3,000-square-mile lake in western Montana. The ice dam eventually failed and caused a month-long flood that unleashed 30 million cubic yards of water per second down the Columbia, forty times the amount of water that currently flows down the mighty Mississippi. The process repeated itself many times there and elsewhere. These floods peeled away vegetation, soil, and bedrock, carving out entirely new landscapes.

Ponderosa pine, which had migrated completely out of the Rocky Mountains, slowly returned and now dominates the Colorado Plateau. Douglas fir had taken refuge during the ice age in small, low, protected valleys and probably along the coastal shelf in Southern California that is now underwater. It migrated north to its current range. Ocean currents might have buffered the Californian coast and protected the giant redwoods. Unlike most North American trees, this species has grown in the same area for as many as 5 million years.

The variously paced species migrations sputtered in fits and starts, making numerous advances and retreats. For example, the climate warmed rapidly and trees moved quickly between 12,000 and 10,000 years ago. Then, approximately 9,000 years ago, the climate became warmer and drier. Precipitation dropped by 25 percent, stressing trees and slowing migration. The warming trend reversed about 5,000 years ago, creating a mini–ice age that benefited some trees and forced others to again retreat south from drought and cold. Northward migration sped up again as the climate warmed from 900 to 1300 AD due to increased sun activity. Temperatures dropped again in 1450 during the "little ice age," which ended around 1850, and again forced some species to retreat southward and down in elevation.

At the beginning of the ice's most recent major retreat, 10,000 to 15,000 years ago, there were more animals, more large animals, and larger animals than at any time since. Charismatic large mammals such as woolly mammoths, mastodons, giant beavers, and saber-toothed tigers vanished from North America rather quickly after this time, perhaps under pressures of human hunting, perhaps because their grazing habitat was replaced by dense deciduous forests in the East and drier savanna conditions in the Midwest.

This brief environmental history shows that nature is dynamic. Change is the norm. Some changes occur slowly, as is the case with shifting tectonic plates, while other changes are swift, as is the case when ice dams release dramatic floods, volcanoes erupt, or asteroids impact Earth. Just as climate change alters species migration, most changes ripple through the web of life, causing other changes. Like a flag whipped by the wind, nature is an undulating fabric in a constant state of flux—each species and organism connects to many others and holds the flag together, but the winds of evolution, climate, and plate tectonics give the flag its shape at any particular moment. Dramatic migration of forests, earth-gouging glaciers, changing sea levels, and shifting river courses all occurred during the geological blink of an eye when humans were present. These changes have not stopped. How can we justify the nature present at one point in time, say, 1491, as being better than another when we know that many natures have existed and many more can exist? As the book progresses, it will become increasingly obvious that deciding which natures should exist requires imposing human values for which natures we want to exist.

NATURALLY SELECTED

Darwin based his theory of evolution on two fundamental observations. First, he observed a great deal of variation among members of the same species. This variability is easy to see in the familiar and faithful canine. Dogs exhibit an enormous range of sizes, colors, shapes, and temperaments. Second, Darwin observed that organisms reproduce at different rates. Evolution occurs, he concluded, because some individuals have variations making them better able to find resources, attract mates, or otherwise generate more offspring. Competition for resources and mates is especially keen among members of the same species because these individuals live in the same ecological niche, using the same methods of reproduction and resource acquisition. The notion of intense competition evokes images of a "nature red in tooth and claw" where only the fittest survive. However, reproductive advantage is a more accurate characterization of the mechanism of evolution; the genes producing the most genes win. Competition for resources is only one test; reproduction is the ultimate measure of success.

"Natural selection" describes the primary mechanism of evolution. New variants of life are tested by current environmental conditions in a race against similar organisms to produce more offspring. Subtle variations occur in each new generation. The variants producing more offspring eventually outnumber and replace other variants. Through this incremental process, small variations are tested, selected, and accumulated over generations. And through this slow, methodical process, the first replicating strand of DNA-like molecules swimming in the oceans some 4 billion years ago evolved into the sophisticated and complex life-forms thriving in all reaches of the globe.[3]

In Darwin's time, it was common sense and unchallenged dogma that nature was perfect as created by God. Changes in creation, if they occurred at all, were intentional efforts by the Creator to improve creation. Darwin challenged this dogma by contending that not only had nature changed, but the changes were unintentional. Darwin did not know the mechanism causing the variations; he just assumed that it was random. We now know that life's variations result from changes in chemical sequences of complex deoxyribonucleic acid molecules (DNA), a process we refer to as gene mutation. Mutations occur whenever DNA fails to replicate itself exactly, often because of some chance encounter with an energy source that disrupts a chemical bond—such as would result from random bursts of solar radiation, chance encounters with reactive chemicals, or unpredictable changes in temperature. Of course, sexual reproduction also creates new genetic information by combining DNA from both parents.

DNA is the means by which evolution passes information forward to future generations. It contains the design instructions for constructing *and* operating an organism. A DNA molecule is much more than a blueprint whose job ceases upon completion of construction. It is an information processor that continually responds to environmental situations by building other molecules, which build still other molecules and perform the functions of life. Subtle changes to DNA's structure can change the structure of the organism and the means by which the organism responds to environmental stimuli.

Evolution Is Random, Not Intentional

Evolution lacks foresight. It does not see an opportunity and respond to it. Rather, random mutations of DNA create random changes to an organism's form or function. Some of these changes grant advantages that get passed on to the next generation. Many more changes do not grant advantages and perish in tests of natural selection. We call the change that grants the advantage an "adaptation." Given the benefit of hindsight, these adaptations look like logical and intentional responses to local conditions. It is tempting to suggest, for example, that giraffes evolved longer necks *so* they could reach higher into the trees for food, or that humans grew bigger brains *so* they could outthink other predators, or that flowers developed relationships with insects *so* they could reproduce more effectively than plants using windblown pollen, or that the male peacock's tail was *intended* to better attract mates. These adaptations do illustrate how evolution exquisitely fits an organism to its situation, but they do not imply intentionality. Evolution has no forward-looking capability. Only when we use hindsight can we can see why a particular random variant was likely to succeed. The power of evolution comes not from its ability to envision solutions to environmental problems but from its never-ending ability to create and test alternative solutions that become the new variants of life.

For example, 400 million years ago (or so), the large landmass that was to become today's separate continents was fracturing and moving northward, creating new habitats and climates for life to explore. The area of edge habitat, where land and water interweave, expanded dramatically. Amphibians evolved from fish and spread to take advantage of these new habitats. Drought would have periodically shrunk areas of the vast swamps, lakes, and inland seas into small pools. Amphibians' survival advantage over fish was their ability to move short distances on dry land and seek new sources of water. They remained dependent upon the moist conditions that dominated the era because these early vertebrates needed ready access to water for respiration and because their eggs were laid, fertilized, and hatched underwater.

Fifty million years later (or so), reptiles emerged through countless incremental genetic mutations, each variation exposed to countless tests of natural selection. Their water-impervious skin, thick-shelled eggs, and lungs would allow them to survive on dry land during all phases of their life cycle. Reptiles possessed little advantage over amphibians in water-rich conditions. But these conditions would not persist. The supercontinent continued its move north, nearer the equator. Heat and landform changes dried swamps and drained inland seas. Reptiles had a survival advantage in these drier conditions. They expanded their range dramatically into newly created and recently vacated niches.

Reptiles were not necessarily better than amphibians when wet conditions prevailed, at least they did not (and still don't) dominate in the moist habitats to which evolution had finely tuned the amphibians. Instead, reptiles rose to dominance when newly created conditions let them utilize potentials that the amphibians did not possess. Evolution, as this example illustrates, is not intentional; rather, it is an opportunistic response to chance and change.

Evolutionary change also is prompted by mutations that stumble onto cooperative, symbiotic, mutually beneficial relationships. Some of these accidental relationships created dramatically new solutions to the challenges of life. A relationship of importance to us all is the cooperation occurring within our cells. Eukaryotic cells are familiar to grade school biology students; they contain a cell wall, nucleus, mitochondria, and other organelles that perform cellular housekeeping. They are the building blocks of plants, animals, and other complex life-forms we know today.

Eukaryotic cells evolved from earlier evolved prokaryotic cells. They more efficiently organize various housekeeping functions and thus are better able to make specialized contributions required by complicated life-forms. These functions are performed by self-contained organelles. The mitochondria organelle, for example, converts complex carbon compounds into carbon dioxide, water, and, importantly, energy. The energy gets trapped inside ATP molecules, which are excreted from the organelle. In essence, what mitochondria produce as waste becomes the fuel that powers cells.

How did this cooperative relationship come to be? Current thinking suggests that many prokaryote cells probably "swallowed" mitochondria-like bacteria and digested them as a source of food. This predator-prey relationship likely occurred billions of times, perhaps for millions of years. Occasionally, rather than consume the mitochondria, the prokaryote cell instead consumed its waste. The prokaryote supplied the mitochondria with the nutrients and safety it needed, while the mitochondria provided the prokaryote with the fuel it needed. Both were better off and outcompeted other prokaryote cells and mitochondria that continued in their adversarial predator-prey relationships. However, the new hybrid cell could not reproduce itself.

After who knows how many billions of cells and mitochondria stumbling onto this symbiotic relationship, other mutations stumbled onto mechanisms that linked their reproductive timing. This linked reproduction cemented a partnership that enabled evolution to explore whole new forms of life. Eukaryotic cells are the building blocks of most of the life-forms we recognize: plants, animals, fungi, seaweed, and amoebas. To this day, the DNA present in the nucleus of a eukaryotic cell is independent of the DNA in the mitochondria. For example, the DNA in the nuclei of the cells of your fingers holding this book is the product of your mother's and father's genes. But the DNA in the mitochondria of those same cells comes only from your mother's mitochondria.

Countless other "accidents" make up the story of evolution. Bottom line: randomly caused mutations create new variants of life, which are tested in environmental conditions that are constantly changing due to unpredictable if not random factors. The creative potential of evolution is in its ability to generate and test countless variations, not in its ability to intentionally design solutions to environmental challenges.

Humans Are Not Special

One implication of the randomness of evolution is that humans are fortuitous accidents, lucky to have evolved. If we could somehow go back several billions years to the beginning of life on Earth and let evolution unfold again, it is incredibly *un*likely that evolution would follow the same trajectory we find in the fossil record. It is extremely improbable that the exact same gene mutations would occur at the exact same time, if at all, to produce the same life variants to be tested by the precise environmental conditions produced by random events such as plate tectonics, climate change, and asteroid impacts. While we humans might think ourselves the stars in the movie of life, we probably would not get *any* role in the remake.[4]

The randomness of evolution does not mean that all life-forms are equally possible. If we could restart life and observe the new life-forms that evolve, we would likely recognize many fundamental properties. Properties such as intelligence, mobility, and sensory awareness may still evolve because they provide survival advantages for the general conditions that characterize Earth. Light, sound, texture, heat, gravity, magnetism, vibration, and chemicals are just some of the media present on Earth by which organisms become aware of their environment and use to find food, shelter, and mates. Mobility, likewise, provides selective advantages. It helps organisms respond to the opportunities and threats identified by sensory organs. Intelligence to respond flexibly and work cooperatively also offers obvious survival advantages.

There are not an unlimited number of ways to perform the tasks required for life on Earth. Therefore, it is likely that alternative evolutionary trajectories would converge on solutions that look similar to those that are successful in species alive today. The species would be different, but the functions they perform might be similar. The randomness in evolution only means that the particular life-forms we know today—primates, beetles, oaks, amoebas—probably would not exist in forms we recognize.

Taking this argument further, one could argue that evolution does show some direction or progress. Evolution does not lead, inevitably, to humans specifically, but perhaps it does flow toward ever-higher levels of ability to assess, organize, and respond to environmental information. Organisms with more of these abilities are better able to survive and reproduce in the environmental conditions present on Earth.[5]

Darwinian evolution demoted humanity. We are accidents of history. We learn from evolution that humans are not the apex of evolution, or in any other way a special product of evolution. Instead we are fellow voyagers in a remarkable odyssey of evolution, connected to all life on Earth by evolutionary history and process.[6] We share with other species natural selection as the judge and jury that tested and created all life's variations. We have much in common with the rest of life, including our good fortune.

Organizing Biodiversity

Despite being demoted by Darwin, we still tend to bias our understandings of nature in ways that inflate our evolutionary importance and inevitability. We project these biases on to efforts to organize and explain species diversity.[7] The ladder and the tree are two organizational metaphors that remain deeply rooted in popular understandings of biodiversity and evolution. Not surprisingly, both metaphors suggest that humans merit a special place nearest the apex of evolution, superior to other types of life.

The ladder metaphor arranges the diversity of life along a linear continuum according to complexity, from early bacteria through single-celled organisms, plants, fish, amphibians, reptiles, birds, and mammals—with humans on the ladder's top rung. The ladder metaphor suggests that life-forms located nearer the top of the ladder are somehow superior to those below it. With a little honest introspection, humans must admit that this assumption is biased. Very simple organisms—bacteria, for instance—are incredibly robust and well suited for the niches they occupy. Their evolutionary success does not require complexity. Why should they be located *below* more complex species? The ladder metaphor is useful only to satisfy human egos by placing us at the top of evolutionary output.

The tree is another metaphor that organizes species. It also suggests that evolution is progressive and humans are superior. Life starts from a common simple ancestor located at the base of the tree and gradually diversifies along numerous branches as life on Earth is tested and "improved." Many branches terminate early, but a few continue outward and upward. The edges of the tree represent the most perfected life-forms yet evolved along that branch of evolution. They are the culmination of adaptations tested by natural selection along a possible evolutionary trajectory. The interior and terminated branches are, by implication, inferior. "Better" species are displayed farther from the origin. The branch reaching farthest from the primitive ancestor contains mammals; its topmost bud, not surprisingly, is *Homo sapiens*. But, as this chapter attempts to make clear, the most recently evolved species are not necessarily *better* than previously evolved organisms; instead it is more accurate to conclude that the most recently evolved are temporarily better suited to current conditions, which will change. Another problem with placing humans at the top of the ladder or tree is that humans are not the most recently evolved. Bonobos—smallish, highly sexed primates living in Congo Basin rain forests—more recently differentiated from other primates than did humans.

A contrasting metaphor for organizing biodiversity is a sprawling, un-kempt, vigorous shrub such as the juniper bush. No single branch dominates this formless bush; rather, branches twist and turn in countless directions. They continue to grow and change and branch in search of sunlight. Occasionally an environmental event (ice, fire, climate change, groundskeeper with hedge trimmers) trims off a branch or two. The strength of this bush, however, comes from its trunk, which contains the processes and mecha-nisms of life that formed early and proved durable. Some inner branches are thick with many trajectories of evolution; others are thin. Some branches are long; others stop abruptly. Some sprout green leaves; others do not. Which branches flourish is more the result of a lottery than a plan. Chance conditions nip some buds and leave others to sprout. The living tissue inside and on the periphery of the bush represents the species present today. One of these branches happens to contain mammals, its location indiscernible from the other branches. Distance from the trunk signifies amount of change, not quality of change. No leaf on the bush is more important than another; no species is better than another.

Species and Biodiversity Are Not Essential Categories

Species are socially and politically important biological entities that we pro-tect with the Endangered Species Act. Species classification systems orga-nize our understanding of Earth's life-forms and conceptually differentiate humans from these other forms. Yet species are neither static nor absolute entities. For at least five hundred years, taxonomists have been refining def-initions of species and still no consensus exists! If taxonomists can't decide what a species is, then how can society defend environmental policies that allocate scarce resources for species preservation?[8]

A review of two competing taxonomic systems illustrates the lack of consensus. Taxonomy, in brief, is the science of classifying individual organisms into classes of larger units (i.e., species) according to the similarities and differences in characteristics such as body shape, size, color, habitat, interbreeding, genetic code, and evolutionary history. Individual organisms can end up being classified in different species categories depending on which taxonomic system is used. A famous and popular classification scheme was proposed by Linnaeus in the eighteenth century. He organized plant species according to observable morphological characteristics such as the position and number of sexual organs (stamens and pistils) and used his understanding of marriage among humans to organize individual species into classes: monogamous species with one husband, one wife were placed in one class; species where females had multiple husbands were in another class; species where males had concubines in yet another class; and so on. His system revolutionized taxonomy because it offered an easy and effective way to standardize naming and identification at a time when people were using different names to refer to the same species. Remnants of his influential system remain embedded in the contemporary classification system, which organizes organisms into a hierarchy starting with phylum and cascading down through class, order, family, genus, and species.[9]

An alternative classification system emphasizes a species' evolutionary history rather than its appearance. The splintering of parent species into new species produces a treelike branching pattern showing evolutionary sequence. As you might expect, there is debate about criteria that differentiate new species from their parents, just as there is debate about the characteristics that differentiate current species from one another. Schemes such as Linnaeus's that use observable characteristics produce different species categories and different organizational structures than do schemes using evolutionary trajectories such as PhyloCode.[10] Which one is correct?

It is tempting to suggest that debates about classification are being overstated: just a tempest in an academic teapot. It is obvious even to lay observers that humans, chimpanzees, lions, tigers, bears, and white oaks differ from one another. But most of the world's proposed species are *not* so easily distinguished by the human eye and intuition. For example, most of us are familiar with lightning bugs that add sparkle to summer evenings. However, few of us can distinguish among the nearly two thousand different species of lightning bugs and thus rely on criteria developed by the taxonomic sciences to differentiate one species from another. Enormous species diversity exists within beetles, barnacles, and bacteria—perhaps tens of thousands of variants. Most of us are unaware of these subtle differences, yet their classification into separate species categories influences estimates of biodiversity and invokes sanctions of the Endangered Species Act, both of which can affect land-use management and policy. Debates over species categories should not devalue efforts to conserve species. We should not stop protecting whales while we quibble over differences among beetles. However, the classification debates are relevant to policies and public understandings about general environmental qualities such as biodiversity.

All Life on Earth Is Related

Evolution teaches us that humans are deeply embedded in and connected to the rest of life on Earth, an incredible ancestry of which we can be proud. Most life on Earth possesses similar genetic and biochemical properties. All life assembles proteins out of the same amino acids and shares the same protein-synthesizing machinery—DNA and RNA. Humans and other life-forms on Earth share most of the same cellular mechanisms and many of the same genes; some estimates suggest that about 95 percent, 74 percent, and 47 percent of human DNA sequences are the same as those in chimpanzees, earthworms, and yeast, respectively. And, obvious to the naked eye, most animals share a basic body plan evolved some 500 million years ago: head, legs, tail, thorax, and left-right symmetry.

Evolution most directly connects us to apes. For reasons unclear, the human species began a separate evolutionary journey from other primates about 5 to 6 million years ago. Another primate, the bonobo, began evolving from other primates into a separate species even more recently than did humans, perhaps just 2 or 3 million years ago. The human primate evolutionary tree includes *Australopithecus afarensis, Homo habilis,* and *Homo erectus.* Very recently, just one or two hundred thousand years ago (only five to ten thousand generations), humans evolved to be physiologically indistinguishable from us living today (*Homo sapiens sapiens*). Humans share remarkable similarity with other primates (which also include bonobos, chimpanzees, gorillas, and orangutans) in our genetic makeup, physiology, and ontogeny. The differences that do exist are matters of degree. Research reviewed later (in chapter 10, "Human Nature") suggests that chimps use language and tools, are self-aware, and can manipulate abstract thoughts. From evolution we learn that there exist deep connections among all life on Earth.

Evolution Is Cooperative and Competitive

The fittest that survive are not always the most competitive: sometimes they are the ones that developed cooperative relationships. Examples of cooperation and symbiosis abound. We are familiar with the charismatic tick bird living on the backs of rhinoceroses. The bird enhances the rhino's survival chances by feeding off the ticks that infect the rhino's skin and by noisily warning the rhino of danger. In turn, the rhino's bulk and mobility protect the tick bird from predators while providing it a source of food. We are also familiar with the less charismatic bacteria such as E. coli that live in our intestines and help with our digestive processes—healthy humans host more bacteria in their digestive tracts than we have cells of our own. Many of these bacteria are anaerobic; they die when exposed to air—literally burned by reactive oxygen. We provide them a moist, acidic, oxygen-poor, temperature-controlled setting ideal for their reproduction. They digest foods we cannot and excrete sugars and proteins we can absorb.

Another example of symbiosis important to humans, and one that puts an interesting twist on evolution, is the role we play in selecting the mutations in plants and animals that get passed on to future generations. One could argue that this process is not evolution but rather domestication because variations are not "naturally" selected but rather "artificially" directed by humans. Such an argument misses the bigger point: both humans and the organisms they domesticate benefit by greater reproductive success (i.e., both gain the evolutionary advantage of producing more genes).

Several thousand years ago when the continents were shaped and located much as they are today but the climate was considerably different, humans discovered agriculture in the area now known as the Middle East (we would do so elsewhere and at different times, but this example is sufficient for the point to be made here).[11] Our ancestors discovered long ago that wild grains were both tasty and nutritious. During seasonal migrations, hunter-gatherers returned to locations where these treasures grew, much as modern humans return to wild raspberry and blueberry patches season after season. The largest seed heads, located at eye level and easily separated from the stalk, were the first and most likely to be picked. These harvests were stored for brief periods at seasonal camps, located near easily accessible water for drinking and flat terrain for sleeping.

People would find, returning to the same sites year after year, new stands of their treasured wild foods growing in their previous piles of waste and defecation. Once it was discovered that planted seeds sprouted to produce additional food, it was a short step to the development of a mutually beneficial relationship between people and these plants—what we now call agriculture. Domesticated plants conveyed great advantage to humans, providing valuable nutrition and a measure of control over an otherwise harsh, variable, and unforgiving environment. In return for this advantage, humans clear away other vegetation, divert water, apply fertilizer, and otherwise do whatever we can to ensure that the plants and their genes reproduce.

Seed heads that did not explode when touched and did not drop to the ground before ripening but remained on the stem for humans to find and harvest were more likely to be picked and planted the next year. Seeds that were larger, tastier, and more successfully stored for long periods of time without rotting were given priority when planting. Over centuries these practices domesticated plants such as wheat, rice, corn, and apples. The genes producing the attributes desired by humans became more prevalent: large edible seeds and fruit, easy collection, durable storage, and so on. These domesticated species, after centuries of human selection, now bear little resemblance to their wild cousins.

Social Darwinism

Evolution is an organizing principle for all biological sciences. Every theory and finding must make sense within the context of evolution. This tremendous explanatory power makes it tempting to look to biological evolution for lessons relevant in social situations. But one of the repeated warnings in this book is to be wary of the temptation to extract *ought* from *is*. The *ought*s we select typically mask values and disguise agendas. Social Darwinism provides a particularly telling example of the dark side of extracting *ought* from *is*. Social Darwinists have used the supposed objective principles of evolution to justify exploitative and oppressive social policies.[12]

Social Darwinism starts by (incorrectly) assuming that competition—survival of the fittest—is the only mechanism by which evolutionary fitness is selected (even though, as illustrated above, mutualism, symbiosis, and other forms of cooperation also promote reproductive success). It extends the ideal of competition to social situations. If competition is unquestionably "good" in biological evolution, then, social Darwinism assumes, it can also serve as the criterion to select cultures, languages, businesses, ideas, and other social phenomena. In other words, all social phenomena should be perfected through competition. The struggle for existence in a highly competitive social environment produces winners, who are in turn deemed good because they survived the competition. Therefore, a society fostering competition is "good" because it consists of all these winners. Nothing should interfere with this competition.

Social Darwinism thus discourages interventionist government and social charity that direct resources to the needy. The rich should become richer and the powerful should become more powerful as they exploit their competitive advantages and pass these advantages on to their descendants. Human society will improve if competition weeds out the weak. Social

Darwinian-type arguments have been used to justify racism. The races enjoying more socioeconomic power were, supposedly, more evolved and should outcompete the less evolved. White Europeans, not surprisingly, placed their culture at the apex of the cultural evolution ladder, with other cultures arrayed below. Contemporary science contradicts this myth. Genetic differences among the so-called races of the human species are rather small, actually smaller than the genetic variability found within a race.

Eugenics is one of the better-known examples of social Darwinism. It deals with the improvement of human qualities through selective breeding for hereditary characteristics. Positive eugenics identifies and tries to pass on desirable characteristics, while negative eugenics tries to reduce undesirable characteristics. There were many social factors that gave rise to this movement, but eugenics was at least partly motivated by the successes of stock breeding at improving animal productivity as well as by the social Darwinist concerns that welfare programs were interfering with the natural selection of humans, allowing "inferior" people to reproduce and therefore degrading human potential. Social traits of white upper- and middle-class Protestants were defined as "normal" and treated as ideals to which others were compared. Deviations from these traits were given labels such as "lame," "criminal," "feebleminded," and "poor." Social programs attempted to remove these people from the gene pool by preventing marriage and forcing sterilization. Social Darwinism, therefore, encouraged the assumption that people of lower socioeconomic status were biologically inferior rather than lacking in educational, social, or economic opportunities.

Eugenics combined with health and social hygiene concerns produced programs that sterilized people with ailments and susceptibilities deemed undesirable, such as low IQ, deafness, and blindness. Nationalism plus eugenics produced calls to stem immigration, which was swelling the numbers of the "unfit." Nazism used eugenics to justify the sterilization of alcoholics, homosexuals, unwed mothers, and others with "moral weakness"; the "mercy-killing" of handicapped children and adults; and the creation of death camps for Jews, Gypsies, and Slavs. Worldwide, millions of people have been killed or sterilized because they were considered to be genetically inferior or even nonhuman. Even in the United States, with its emphasis on individual rights and personal freedom notwithstanding, hundreds of thousands of sterilizations were sanctioned using the logic of eugenics. Indiana passed the first statewide sterilization law in 1907, a practice that continued in some states well into the 1970s, long after the science of eugenics had been discredited.[13]

Social Darwinism illustrates the dangers of using supposed lessons of nature to justify social policy. It selectively emphasizes competition as *the* lesson of evolution because doing so conveniently justified oppressive social polices that promoted agendas of people in power. Social Darwinism declined in popularity by the mid-1900s after closer inspection by biologists, philosophers, and social theorists revealed misinterpretations of evolutionary theory, misunderstanding of heredity, naïveté about social processes, and aggressive social agendas masquerading behind the veil of science. Biological and psychological studies increasingly show that both nature and nurture determine human qualities.[14] Human nature, it seems, is very sensitive to human culture. Moreover, after World War II, public tolerance waned for movements associated with Nazism and racism.

CONCLUSION

Random variations persistently tested over eons produce astonishing, beautiful, and diverse solutions to the challenges of living on Earth. That process—evolution—continues. It is creative. It and human intelligence are the two most powerful forces on Earth. They are the only known processes that are flexible, adaptive, and capable of generating vast alternatives to ever-changing environmental challenges. Evolution provides humans a worthy partner. It has verve, integrity, creativity, and power—qualities we respect. It is something beyond ourselves toward which we can direct respect and love. We should be cautious of exporting lessons from evolution about how to organize human society, but there seems good reason to respect evolution's creative potential and our evolutionary connections to all life on Earth.

* 4 *
Ecological Nature

Holism • Mechanical
Disturbance • Forests • Farms • Natives
Exotics • Ethics • Leopold

A BALANCED ECOLOGICAL WEB?

"The balance of nature" is a phrase that rolls off tongues during environmental debates. Some speakers invoke it to argue against human intervention that might disrupt Earth's fragile ecology. Other speakers use it to defend the opposite conclusion, arguing that nature's robust and redundant systems can withstand aggressive resource exploitation. These speakers appeal to deep-seated cultural myths about the harmony, health, integrity, and stability inherent in the interconnected web of nature. But is there really a balance to nature? Does nature know best? The topic is hotly debated.

Consider the interconnectedness and dynamism of nature as illustrated by the journey through the biosphere of a hypothetical particle, X. Eons ago X settled out of the dinosaur food chain and onto the bottom of an ancient ocean. Time and pressure imprisoned X in sedimentary rock. Climate change dried the ocean, and plate tectonics raised a mountain. Weather and roots made soil from the rock. Eventually X moved through the porous membranes of a root hair, through countless tree cells, and into another food chain. In a very few years, X traveled from acorn to deer to human. Through cycles of death and decay, X again became soil, only to be sucked up by another root, turned into another plant, eaten by another chain of animals, and returned again and again to the soil. X traveled through cycles of life, death, and decay in countless organisms, but stayed in pretty much the same geographic location despite dramatically different conditions wrought by climate, flood, fire, disease, and evolution. Eventually X washed into a stream and returned to the ocean, where it cycled through sea life until imprisoned again in sediment that will become the bedrock of future forests. No balance exists to define X's ideal location. X has flowed through natures dominated by volcanoes, oceans, plants, dinosaurs, and humans. It will flow through many other natures. Flow and interconnectivity are the lessons of this parable, not balance.[1]

The state of nature when X is a tree is not necessarily better or worse than the state of nature when X is bedrock or a dinosaur. Ecological systems, thus, are not balanced in any literal sense; rather, they are porous and dynamic. X's journey shows that even organisms—which to the human eye seem discrete and self-contained—are really nothing more than dynamic flows of matter and energy (i.e., you are shedding and assembling countless cells and molecules as you read this parenthetical comment). For confirmation of the lack of balance in nature, one needs to look no further than the dramatic and continuing changes described in the chapter on evolution. Many previous natural conditions existed, and many more will follow. Change, not static balance, is the lesson of geology. The final nail in the coffin of balance comes from anthropology. As reviewed in chapter 2, human activities are inextricably intertwined with environmental conditions and have been for some time. Nature cannot be in balance because human culture is dynamic.[2]

Appeals to nature's balance are misplaced attempts to extract *ought* from *is*, a false hope that ecological science can identify objective management goals. If ecological systems were balanced, it would simplify environmental politics, science, and management. The science of ecology would strive to define balanced systems and how far they can be modified without damage. The politics of balanced systems would be simplified because science would objectively define acceptable environmental quality. Managers of a balanced system would simply strive to control dangerous disruptions to the balanced ideal. Unfortunately, nothing in environmental management is simple. As X's journey illustrates, nature is not balanced and ecological science cannot tell us where X *ought* to be. The absence of ecological balance as an ideal means that the goals of environmental management are subjective. Other criteria, such as community values, must be used to define the desired conditions that guide management. Important efforts to integrate these values into ecological theory have focused on resilience, integrity, and health. Subsequent chapters strive to provide an understanding and appreciation of these values.

THE ORGANIZATION OF NATURE:
WHOLES, MACHINES, HIERARCHIES

The ecological web is another common and influential metaphor in environmental decision making, and deservedly so. The web metaphor does not imply balance; rather, it implies interconnectivity. We cannot put our fingers into the ecological web without causing vibrations and changes elsewhere, many of which loop back to affect us. Interconnections give the web its power and function; individual strands have little utility or durability on their own. Webs can and do change shape: strands of a spider's web break, disappear, and combine with others. These events change the web's function and shape, but it is still a web. Though, at some degree of change, the web not only loses shape; it also loses resilience and functionality. The web collapses on itself, sets adrift on a stiff breeze, and falls to the ground, where its parts disintegrate and scatter.

Ecological science confirms that multiple, complex weblike interconnections exist in nature. In fact, ecology can be characterized as a study of these interactions. At issue is not whether interactions exist, but the conceptual models used to interpret them. There exist three dominant models on which this section will focus: mechanistic, holistic, and dynamic equilibrium.[3]

A mechanistic analysis of nature's web emphasizes the individual strands or parts. These parts can be rearranged and exchanged to improve the machine. The character of the machine does not matter as much as its outputs. Parts are replaced and rearranged in creative ways to produce new machines for producing desired outcomes, such as economic goods and aesthetic services. The parts of a car, for example, can be replaced or reassembled to make the car faster, sportier, safer, or more fuel efficient—whatever outcomes people desire. Tires, engines, paint, windshields, oil, and spark plugs are replaced to make the car "good as new." Only if one desires to preserve an authentic, historic, original design does one worry about which parts are used for replacement, and in these cases one is treating the car as a whole with an identity rather than a machine providing transportation.

The mechanistic interpretation of nature reflects modernist thinking traceable to Descartes' philosophy and Newton's physics, and dovetails neatly with the technological and economic views of nature discussed in the next two chapters. According to these views, human ingenuity, properly applied, can improve upon nature's inefficiencies. The machine metaphor also complements some religious teachings discussed in chapter 9, which suggest that humans have dominion over nature and that science provides a means to finish God's creation. Ecology, economics, technology, and religion therefore can be interpreted in ways that support a mechanistic view in which the parts and processes of nature are routinely and systematically manipulated to produce outcomes valued by humans.

Our social systems require a steady flow of resources to maintain economic efficiency, avoid social disruptions, and satisfy static agency and legal standards. Managers, therefore, are under social pressures to sustain constant supplies of wood, water, forage, and other resources. Nature is treated like a machine to control or eliminate insect infestations, wildfire, weather, and other potential disruptions. Outputs such as boards, salmon, cattle, and cotton are sustained at levels optimal for economic profit and social stability.

Holism contrasts dramatically with the mechanistic view of nature. Ecological wholes are assumed to emerge from the web of interactions. These ecological wholes are supra-organisms, analogous to individual organisms (e.g., a person, a deer). They possess valued properties such as health, integrity, illness, resistance to disturbances, and capacity for self-renewal.[4]

The whole is greater than the sum of the parts. The whole has value in and of itself, and replacement or damage to the parts damages the integrity of the whole. An analogy to a living person may help illustrate the implications of conceptualizing natural units as wholes. I recognize you as a distinct whole. You have distinct boundaries and properties that demand attention and respect. You are unique, functioning, living, breathing, thinking, and feeling. You have a will to live that can be seen in your actions to sustain and protect yourself. You have a clear beginning and ending—you were born and you will die. It might be possible to delete or add a few parts without damaging your essence: for example, a finger, a leg, a kidney, glasses, hair, a cane, drugs, or a hip. With a bit of science fiction, we might imagine inserting computers and other artifices into your body. At some point in this thought experiment, you stop being human and become something else because of the additions and deletions, perhaps a cyborg, perhaps a Spider-Man, perhaps a Frankenstein, perhaps something we can't imagine. Your essence has been changed. Philosophers, priests, and science fiction writers debate the existence of this essence, its value, and how much we should worry about changing it. Similarly, environmental ethicists and policy makers debate whether ecological wholes exist and how much we should worry about protecting their integrities. Holism presents a moral dilemma for land managers because their actions can change or damage the integrity of these ecological wholes.

Ecological theories of plant communities and plant succession advanced by Frederic Clements in the early 1900s provide one well-known example of ecological systems interpreted through the lens of holism. Clements argued that plant communities constituted a new form of life, analogous and maybe even superior to an organism. These communities progressed through a series of development stages just like an organism, from birth through to maturity. Specifically, a plant community is born when it colonizes a recently disturbed (i.e., burned) site and juvenile plant combinations emerge, which are followed by maturing species compositions until a "climax" system emerges.

Clements believed that the climax community was a whole existing at "a higher order than an individual geranium, robin, or chimpanzee.... Like them, it is a unified mechanism in which the whole is greater than a sum of its parts and hence it constitutes a new kind of organic being with novel properties." His words leave little doubt that his understanding of nature as an organism was literal as well as symbolic: "The unit of vegetation, the climax formation, is an organic entity. As an organism, the formation arises, grows, matures, and dies. . . . The climax formation is the adult organism, the fully developed community. . . . Succession is the process of reproduction of a formation, and this reproductive process can no more fail to terminate in the adult form in vegetation than it can in the case of the individual."[5]

Clements's assumptions have been roundly critiqued. Historians of science show how they reflect his political and cultural context, which also birthed socialism, nationalism, progressive conservation, and related holistic ideals. Notions of ecosystems as living supra-organisms no longer carry much sway in ecological science, being replaced by systems theory, macroecology, and meta-communities. Revised holism, however, remains an important way to understand nature because unique properties and functions do emerge out of complicated ecological systems, and these properties can be valued as much or more than the parts that created them. Hence, the moral dilemma of managing wholes remains a source of heated debate.

Dynamic equilibrium provides yet another means of understanding the ecological interactions that make up nature's web. Ecological systems are understood to be hierarchical clusters of fragmented and dynamic patches constantly fluctuating due to internal evolution and external perturbation. Change and chaos, not balance, are the norms. Complexity, nonlinearity, and randomness make predictions difficult; no matter how well we understand a system, small random events can result in large unpredictable outcomes. Still, some semblance of equilibrium exists. When viewed from a temporal scale of several centuries and a geographic scale of many acres, it appears that the same or similar ecological parts and processes persist over these intermediate temporal and spatial scales, even after repeated disturbances. Their persistence is a type of equilibrium or resilience to change.

Hierarchy theory explains nature as a set of systems nested inside one another, starting at larger scales of time and space and cascading down to the local and immediate.[6] The nested systems influence each other: impacts course up from smaller to larger systems and down from larger to smaller systems. Changes generally occur more quickly at the smaller scales. For example, parts of a forest change quickly and dramatically due to weather, insects, and fire. Damaged trees die, fall, or are toppled by wind. Fallen trees fuel hot fires that burn off soils. In spite of all this perturbation, the major functions and components of large contiguous natural systems persist because neighboring undisturbed areas serve as temporary refuges that can reseed the disturbed areas with plants, animals, and processes. The species composition and ecological processes occurring across the larger landscape therefore have resilience despite the repeated occurrence of smaller disturbances. In fact, a pattern of repeated disturbance might be essential to maintaining some system properties that have become adapted to repeated change.

Changes in the smaller systems are unlikely to affect the larger system's resilience unless many proximate smaller systems change similarly, as might be the case when sprawling suburban areas create forests with understories of swing sets, invasive species change vegetation composition, catastrophic fires sweep through even-aged forests, or insect infestations consume monocultures. When these larger systems change, they can release large amounts of nutrients that will be reorganized; the life in the smaller patches will explore and colonize new habitat. The invasive, aggressive, opportunistic species will be freed from the feedback and control previously provided by the larger system in which they were embedded. Over time, a new (and probably different) hierarchical system of interdependencies will emerge and begin to organize, recycle, and conserve resources until (or if) a new flip of the system is triggered and stored nutrients are again released. Thus, any attempt to describe the stability or resilience of nature depends upon the temporal and geographic scale being examined. A hierarchical perspective suggests managers should anticipate the impacts of their actions across multiple scales so as to sustain system resilience and prevent catastrophic system reorganization.

IMPLICATIONS OF DIFFERENT
UNDERSTANDINGS OF NATURE

The three different understandings of ecological interrelationships reviewed above—mechanistic, holistic, and dynamic equilibrium—are used to justify very different management strategies.

The Mechanical Farm

The tremendous productivity of agriculture provides a fertile example of mechanistic nature (and the compatibility of the mechanical model with the technological and economic understanding of nature discussed in chapters 5 and 6). Industrial agriculture strives to maximize the value of outputs (crops) while minimizing the value of inputs (seed, labor) by improving efficiency of the machine (farm ecosystem). Farms could be managed holistically, as partially closed systems where solar energy and rain are the main inputs and the only outputs taken to market are the result of plants converting that rain and sun into grain. Each year enough manure and seeds are saved to fuel and sustain the productive cycle. Farms managed mechanistically, in contrast, insert materials and energy at one end and collect outputs at the other.

> Now farmers buy their seed from Pioneer Hybrid Seed Company, their "mules" from Ford Motor Company, the "hay" to feed these "mules" from Exxon, and the "manure" from Union Carbide. Thus farming has changed from a productive process, which originated most of its own inputs and converted them into outputs, to a process that passes materials and energy through from an external supplier to an external buyer.[7]

This mechanistic model has made agriculture among the most efficient industries on the planet. Investing in new technologies and economies of scale allows farmers to produce more crops at a lower cost per acre. Early adopters who improve their efficiencies are able to realize greater profits. Those who don't adopt the technology are bought out by those who do, increasing the economies of scale and profits for those who remain. Society enjoys more food at a lower price as these technologies disperse throughout the market and increase food supply. Adoption of the new technology by more and more competitors means no farmer long enjoys a competitive advantage. The early adopter's profit margins eventually shrink, and farmers seek new technological advances and greater economies of scale. As farming becomes more and more efficient, the prices of farm products continue to drop, sometimes below the costs of production because governments subsidize farmers in order to curry political favor and protect the national food supply.

Mechanical Woods versus Whole Forests

The conception and practice of forestry in the United States illustrate how the mechanistic and holistic models of nature produce very different management policies and natures. Even what we call the "forest" reveals our biases. By calling it "woods," we emphasize building materials produced by trees over all the other aspects of the forest system. "Forest sustainability," defined from a mechanistic perspective, sustains a flow of wood products. A sustained yield of wood is accomplished by taking out all the unnecessary parts, such as "weed" species that compete with desired "crop" species. Pesticides further simplify the forest by removing insects that feed on trees and other forms of life that feed on insects. Ideally, the wood-producing machine would be composed of trees genetically engineered at a nursery to grow fast and straight, planted in well-drained soil that does not erode, and injected with water, nutrients, sunshine, and carbon dioxide as needed. Each tree's location would be carefully controlled so as to maximize its access to air, water, soil, and sunshine without wasting these inputs. Trees planted in long, straight rows speed mechanical harvesting, making the forest one efficient machine from input to output. Trees in this mechanical forest do not reach old age. Older trees, like older people, become less efficient biological machines that stop growing taller and are more likely to get sick and die.

For many years professional forestry advocated the ideal of a "regulated" forest machine. Foresters strove to increase the sustained yield of the nation's wood supply by replacing old, stagnant, diseased trees with young, thriving, healthy ones—hence the desire to clear huge swaths of forests. The ideal regulated mechanical forest would produce a regional landscape resembling a large checkerboard, each square planted in trees of the same age. After harvesting all the old trees in one square, new, genetically improved trees would be planted. The trees would be harvested as soon as annual growth slowed significantly, typically between ten and ninety years, depending on the species and on the desired wood products. This process would repeat itself ad infinitum, producing wood with each rotation of the forest machine.[8]

While such a machine may sustain a steady yield of wood, it might look and feel differently than the forest most people imagine. It is not the forest that is being sustained; rather, it is the flow of wood. Old trees, native species, trees of various ages, meandering pathways, downed trees full of insects, dead snags full of holes and habitat, and variability in species and scenery are replaced by uniformly straight rows, little understory vegetation, domesticated monocultures, and large clear-cuts. The mechanistic forest system is simplified; parts are eliminated and replaced, and concerns about native species and species diversity are ignored. Old-growth forests— with their variability, decay, and slow growth—are considered wasteful and inefficient by comparison.

From the perspective of holism, this mechanized forest would be ruined and raped. It is no longer a forest. It is a plantation, stripped of many parts and lacking control of its most basic natural functions, even the birth and death of trees. From a holistic perspective, native species, species diversity, pollination, natural regeneration, and old growth are the defining qualities that create the character, integrity, or spirit of the whole. Wild processes that create variability in species and decay on the forest floor are viewed as adding positive aesthetic, spiritual, moral, and biological value rather than as wasting productive potential.

From the perspective of dynamic equilibrium, the mechanical forest would be considered frail rather than resilient. Efforts to design a machine that maximizes the flow of wood products would so simplify the forest system that it could be vulnerable to disturbances that periodically and inevitably sweep through whole regions. The hierarchical forest sustains resilience to perturbation because of diversity, variety, and redundancy of ecological function and content across multiple scales of time and space. Harvesting wood from the hierarchical forest occurs on smaller units and only up to the point that it does not jeopardize the resilience of the larger system.

Dynamic Forests

Some attempts at mechanizing nature produced unintended consequences, if not disaster. The case of the Blue Mountains of eastern Oregon and Washington illustrates that dynamic equilibrium is a better conceptualization of some forests.[9] European settlers to this area at the dawn of the 1900s found magnificent forests of towering ponderosa pines averaging three feet in diameter. These deep-rooted, thick-barked trees survived drought, fire, and high winds to live several hundred years. Human efforts to manage the forest like a machine produced perverse results. The first managers removed fire from the landscape and replaced it with logging. Eventually dense stands of shade-tolerant fir trees outcompeted and replaced the more desirable pines. Not only were the pines not regenerating, but intense, unprecedented waves of fire and insect infestations swept through the susceptible firs, devastating both the forest and the regional forest-products economy.

Tremendous resources were deployed to manage the forest machine so as to grow the desired ponderosa forest. Fires were reintroduced when managers realized that shade-intolerant pines required fire to clear out the firs and release stored nutrients into the soil. Logging practices and pesticide applications were redesigned to mimic the effects of fire clearings. But few things worked as planned.

Management efforts that reduced variability created homogeneous conditions that allowed fire, insects, diseases, drought, and other disturbances to spread across whole landscapes. The benefit of hindsight now suggests that variability rather than mechanistic control is needed to produce conditions conducive to ponderosa pines. Periodic cycles of drought, disease, insect infestation, forest succession, and fire combine with soil properties, grazing, nutrient cycles, and hydrology to create a patchwork mosaic necessary for pine regeneration. Mechanistic simplification of the system disrupts the variability under which pines thrive.

Function versus Composition

Land-management policies emphasizing content differ from those emphasizing function.[10] Compositionalists, as the name implies, champion the composition of the ecological web. Entities, such as organisms and species, matter because the parts of functioning systems have value in and of themselves. They have survived tests of time and perform valuable ecological functions, many of which may not be known. Compositionalism, because it emphasizes keeping all the parts, has much in common with holism.

Functionalists, as the name implies, emphasize ecological functions such as flows of energy, water, carbon, phosphorus, and nitrogen. Imagine a large diagram showing solar energy flowing in and being captured by photosynthesis or reflected out into space and forever lost. Plants, animals, soil, and fossil deposits are arranged on the diagrams according to their functions as storage devices for energy and materials. The processes by which energy and materials get captured, produced, stored, consumed, and transferred create the lines in the diagram connecting the parts. Functionalists emphasize protection and management of the functions these lines represent.

A wave provides a classic illustration of the functionalist argument. Gravity hurls water over submerged rocks with such force that the rebounding water creates a large standing wave that evokes fear and respect in canoeists and rafters. There it waits, in the middle of the rapids, ready to tip or swamp your craft, bash your body against the rocks, and hold you submerged in a hydraulic. You have no doubts about its existence and wisely contemplate a portage. However, the wave does not exist as a fixed entity. It consists of an ever-changing collection of water molecules. To understand the wave (and manage or protect it) requires understanding (and managing) the processes that create it.

The functionalist understanding of nature overlaps with the mechanistic view of nature. The parts can be exchanged as long as critical processes remain. For example, some species could be viewed as redundant because other species perform the same ecological functions. Extinction of a redundant species would be of less concern than the extinction of a species that performs unique ecological functions.

The Chesapeake Bay, once one of the world's most productive estuaries, was populated with oysters that filtered all the bay's waters every few days. The oyster population recently declined to less than 2 percent of its historical levels due to parasites, habitat loss, and harvesting. Now the filtration occurs only once every several months, at best. Without filtration, dissolved oxygen content and light penetration are reduced, creating conditions harmful to myriad of valued ecological and social qualities. From a mechanistic or functionalist perspective, it does not matter whether the filtration function is performed by the native oyster, a robust exotic Asian oyster being considered for introduction, or some human-built filtration system. Functionalists would use science to decide whether the exotic oyster adequately replaces other functions performed by the native oyster and would worry that the effects of different dietary needs, reproductive timing, and shell structure will ripple throughout the bay ecosystem. Composition-alists would argue that the emphasis on function ignores vital links of the native oyster to the cultural traditions and natural histories of the bay. The Asian oyster, if introduced, would change the very identity of Chesapeake Bay. Compositionalists also might reject the functionalists' analysis as un-realistic, arguing that science will never know which species can adequately replace all the functions of another because the interconnections among species have been finely tuned by centuries of evolution.

Functionalists and compositionalists can differ dramatically in how they define key indicators of ecological quality such as integrity. Ecological in-tegrity, defined from a compositionalist perspective, is "a balanced, inte-grated, adaptive community of organisms having a species composition, diversity, and functional organization *comparable to that of natural habitat of the region.*"[11] Note the emphasis placed on *natural* habitat of the region. Functionalists would not care which species or functions exist or whether or not they were natural as long as they create a resilient and integrated system. Compositionalists, however, insist that ecological entities (such as genes, species, and assemblages of habitat and populations) must be "nat-ural," "normal," or "native." Typically what makes something natural or normal is that it was produced by evolutionary history rather than by human accident or engineering. Pre-European or prehuman conditions are fre-quently nominated as the benchmarks or ideal states for determining ecological integrity. As the previous chapter on evolving nature suggests, however, natural conditions have changed over time and selecting a non-arbitrary condition as "the" baseline presents serious scientific and political challenges.

Native versus Exotic Species

Exotic, alien, introduced species are viewed with suspicion and fear be-cause they can cause rapid and extensive ecological change.[12] For example, species such as the chestnut blight removed the dominant tree species and changed the entire structure of eastern forests in just a few decades. By 1940 the once-ubiquitous chestnut tree became hard to find. The gypsy moth is currently having the same effect, only this time attacking other species of trees. Kudzu and honeysuckle choke the forests that remain. Zebra mussels outcompete aquatic life in lakes, change ecological interactions among species, and clog municipal water intake pipes. And escaped snakes consume unsuspecting birds on previously snakeless islands such as Hawaii and New Zealand, rapidly extinguishing whole species. The extent and rapidity of these changes evoke fear of all exotics.

Yet some exotics are appreciated for the ecological and agricultural roles they play—bees pollinate crops and yield honey, for example. In spite of myriad attempts, currently no economical means exists to replace the pollination services these bees provide. Many domesticated plants are similarly pervasive and appreciated, lawn grasses and apple trees just two examples among them. The grasses carpeting suburbia are domesticated from wild and imported grasses that tolerate foot traffic and temperature extremes. Apple trees, domesticated from wild cousins in Asia, can be found in forests, farms, and yards almost everywhere.

Environmental fundamentalism can turn discussions about exotics into polarizing shouting matches. Holists and compositionalists see clear distinctions between native and exotics. Natives are good, by definition, because they define the whole. Exotics are bad, by definition, because they ruin the whole's integrity. End of discussion. There is no place for exotics, or people, from this fundamentalist perspective.

The actual debate, of course, is more nuanced. Deciding the value, place, and rights of a species depends upon more than its status as exotic or native. People associate deep aesthetic, spiritual, and moral values with native species. Perhaps people are concerned with equity, believing natives have a longer and thus more authentic claim to exist in their habitat. They may also value each species' long and fascinating history of survival through adaptation, competition, and cooperation. Or they may value each species because it is a creation of God, valued by God; or because it is the moral equivalent to a book in the library of life. Therefore, threatening a native species by introducing an exotic species would be the equivalent to tearing out chapters of the Bible or burning the last copies of classical books from the great libraries of human civilizations. Subsequent chapters on religion, aesthetics, and morality (9, 12, and 13) discuss these qualities in detail.

From a mechanical and functionalist understanding of nature, it does not matter whether a species is exotic or native as long as it efficiently performs the desired ecological functions. The main concern is not what species performs the function, but instead what functions are performed. Arguments against replacing natives with exotics also invoke concerns about humanity's technological prowess. As discussed in chapter 5, "(In)finite Nature," technological optimists believe that human ingenuity can reengineer ecological systems by replacing native species with exotic species or other technologies. Technological skeptics, in contrast, worry that humans will never understand or manage ecological systems well enough to replace native species. They note that little is known about the functions that most species perform or how their collective interactions make ecological systems operate. Thus, technological skeptics, even those subscribing to a mechanical model of nature, might oppose exotics because they doubt the abilities of humans to reengineer ecological systems and prefer the time-tested performance of natives.

Exotics are also a concern from the perspective of dynamic equilibrium. Invasive exotics can change the composition and function of so many smaller patches that the larger ecological hierarchy gets disrupted, pushing large regions into whole new ecological conditions. From this perspective, homogenizing the world's ecosystems with a few species makes all systems less resilient and more vulnerable to dramatic change. Pathogens that develop and destroy a single species used to have a limited global effect because multiple species performed similar ecological functions in multiple smaller patches. Agriculture, monocultures, and urbanization have simplified many ecosystems, so the impacts of adding or removing a species may be more pronounced. Pathogens that previously were geographically isolated because their hosts were geographically isolated in patches can now infect ecosystems all over the world in a relatively short period of time.

ECOLOGICAL LAND ETHICS: TYRANNY OR SUSTAINABILITY?

Aldo Leopold

Aldo Leopold, born in 1887, grew up while the American frontier vanished. He hunted, hiked, explored, and studied wild remnants in the majestic bluffs overlooking the Mississippi River near his home in Burlington, Iowa, and also on Marquette Island in Lake Huron, where his family summered. His family's furniture-manufacturing business was successful enough to send young Leopold to Yale for a college education. He received his master's degree in the brand-new field of forestry, which taught him to see nature through a technical lens, to manage forests as a machine, and to view the products of forests as resources that fueled the economy. His first professional job was with the U.S. Forest Service in New Mexico, where his boyhood love of nature flourished.

At the U.S. Forest Service, Leopold learned firsthand the challenges of managing dynamic and complex nature. His focus gradually shifted to wildlife management, and his scientific study and writing became foundational for the then just emerging science of ecology. Leopold went on to become a professor of wildlife management at the University of Wisconsin and the owner of a farm in Sauk County, Wisconsin, where he established practices of restoration ecology. His famous book, *A Sand County Almanac*, blends ecology, ethics, and poetry, and his astute observations of nature and society have evoked deep passion and reverence in generations of readers. He developed an ecological perspective as an alternative to the economic/mechanistic perspective that then dominated his professional training and contemporary culture.[13]

Leopold looked warily toward a future where nature was entirely subjugated and treated like a machine. Mechanical nature may be efficient, it may fuel the economy, and it might even be sustainable, but it does not inspire a culture Leopold respected. He equated it with "slavery." He acknowledged the remarkable gains in wealth, convenience, safety, health, education, and entertainment enabled through mechanical control of nature but found these an insufficient basis for humanity and community. He worried that nature defined exclusively by efficiency and profit would squeeze out and replace the qualities of life he valued.

Nor was Leopold comfortable with strong holism as an alternative understanding of nature because it defines nature as something independent of humanity. Leopold was not misanthropic. He was a tinkerer, a restorer, and a manager. He treasured the "cultural harvest" and enriched "perception" produced when people *interact* with nature. To him, the purpose of land management was to sustain and improve the "beauty, stability, and integrity of the biotic community," which, in his view, *included* human culture. Humans and nature should be partners or lovers whose union creates a larger, more valuable whole. Leopold promoted ethics and policies that increased the value and sustainability of the union or partnership between humans and nature. He rejected the false choice between holism *or* mechanization because it limits and might ruin the community by forcing a choice between nature and humans, and because it ignores the emergent properties of a thriving human-nature community.[14]

Leopold argued that history shows humans extending the boundaries of the membership of our community and the rights and protections that membership affords. Initially only members of the immediate family or tribe were granted membership. We slowly extended membership to peoples of different regions, nationalities, genders, ethnicities, and religions as we grew to realize our similarities to and dependencies on them. Because Leopold realized our dependency on and connection to the biosphere, he argued that rights and privileges should be extended to nonhuman elements of the biota. His land ethic "simply enlarges the boundaries of the community to include soils, waters, plants, and animals, or collectively: the land." By extending membership to other parts of nature, he gets us to admit that we are part of and dependent on nature. Such an admission bridges the human-nature divide that often polarizes issues into holistic *versus* mechanistic or compositionialist *versus* functionalist dualisms. It shifts the focus from the zero-sum game of humans *versus* nature to the open-ended synergy possible when humans and nature pursue shared interests. Leopold thus sought to change "the role of *Homo sapiens* from conqueror of the land-community to plain member and citizen of it."

It is important to iterate that Leopold's ecological ethic was not anti-human or anti-economic. Leopold clearly includes humans as part of the ecological system. He valued both the material and cultural harvest of nature. In fact, rather than starting with natural systems and extending rights and limitations to humans, as some environmental ethicists do, Leopold started with the human community and extended rights and limitations to the larger biotic community. To avoid charges of misanthropy, which understandably limits the appeal of any land ethic, Leopold advocated an ethical view that unifies rather than separates humans and nature.

Leopold recognized that our work does not stop when the rights, privileges, and responsibilities of membership are extended to nonhumans; rather, it just begins. Community members must negotiate the ideals that define the community and the rules that govern the behavior of its members. Civil dialogue, politics, religion, ritual, and economy are just some of the forums in which these negotiations occur. The real power in Leopold's arguments comes not from extending community membership to nonhuman elements of nature, but from his assignment to community members of the tasks of defining and implementing conditions that create acceptable levels of beauty, integrity, and stability in the community. Community membership, rather than an end in itself, is the mechanism by which rights and expectations are defined and enforced. Membership imposes "a limitation on freedom of action in the struggle for existence"; it imposes "obligations" to limit actions that harm others or degrade the larger community. In other words, Leopold asked us to find ways to show "respect for the community as such."

We still struggle describing and understanding Leopold's land community because it transcends the human-nature dichotomy to include both human civilization and ecological systems. Community ecologists tend to focus only on interactions among "natural" biotic and abiotic features of the environment, while community sociologists tend to focus only on interactions among humans and human institutions. Leopold asks us to overcome these isolating and polarizing tendencies and focus our energies on sustaining his more inclusive, unifying notion of community.[15]

Leopold's definition of community shifts focus from the production of economic resources (i.e., wood, water, recreation) that support the human community to the beauty, integrity, and stability of the larger land community. Thus, Leopold asks us to look up a level or two in the system hierarchy. This different scalar perspective makes Leopold one of the first advocates of the hierarchical understanding that now dominates ecological science. By getting us to focus on community functions, Leopold has us concentrate on what is left in the community rather than what is taken out of it.

Leopold did not deny that humans have dreams, industry, and motivations that change both the land and the community. His calls for stability and integrity were not calls to arrest change or revert to some misanthropic ideal. Rather, he sought the opposite: to sustain the community during its inevitable evolution. The basis of Leopold's land ethic, the core of his arguments, rested on his concern that American society would fail if the land were not sustained. Every culture, every system, is submitted to the test of time. If a community's land-use practices produce ecological collapse, then that community also will collapse, proving it inadequate. His apocalyptic warnings, written at the dawn of World War II, illustrate this concern:

> A harmonious relation to land is more intricate, and of more consequence to civilization, than the historians of its progress seem to realize. Civilization is not, as they often assume, the enslavement of a stable and constant earth. It is a state of mutual and interdependent cooperation between human animals, other animals, plants, and soils, which may be disrupted at any moment by the failure of any of them. Land-despoliation has evicted nations, and can on occasion do it again.[16]

Thriving communities must nurture the will and the means to sustain themselves; they need to provide members not just an acceptable life, but a good life that motivates voluntary restriction of personal freedoms for the greater good of the community. A task of this larger community, then, is to deliberate the social and environmental qualities that sustain the good life and to impose limits on personal freedoms that threaten those qualities.

Policies advocating restraint of personal freedoms are typically suspect in Western cultures founded on personal freedom and autonomy. Thus, both ecology and Leopold have been accused of being fascist and anti-market because individuals must limit individual freedoms in order to protect the good of the ecological whole.[17] For example, if a deer herd grows too large and starts overgrazing the forest, hunting may be required to thin the herd and protect forest regeneration for long-term forest health. Thus, individual deer must give up their lives so that the deer herd may thrive. If exotic species invade an ecosystem and threaten its integrity, then these exotics may have to be sprayed, chopped, or otherwise killed to protect native ecosystem integrity. If fire is required to restore savannas to their presettlement conditions, then deer, rabbits, and trees will be burned to death by prescribed fire. If human economy threatens ecological health, then jobs may be sacrificed to slow growth. If industry pollution damages ecological functions, then expensive mitigation technologies may make some goods and services less affordable. Obviously these trade-offs will be hotly debated in societies that value individual rights and unregulated markets.

To help people appreciate the ecological interconnections that sustain their lives and willingly sacrifice personal freedoms on behalf of the whole biota, Leopold advocated fostering in people a "perceptive faculty" (discussed in more detail in chapter 12: "Aesthetic Nature"). People would more likely develop and practice an ecological ethic if they could appreciate "the incredible intricacies of the plant and animal community—the intrinsic beauty of the organism called America."[18] The ecological aesthetic he described and advocated generates intense and positive emotional reactions to the land. Leopold argued that aesthetic appreciation "begins with the pretty" but deepens with an understanding of ecological and evolutionary sciences. This understanding enriches and deepens people's emotional connection to the land, evoking the feeling of love. We act unselfishly toward people and things we love. We mourn their passing, we crave their company, we sacrifice our well-being for theirs, and we give them things to make them whole, happy, and healthy. When we love the land, we will act unselfishly toward it. The key to promoting an ecological ethic and sustainable behavior in people is therefore appreciating the deeper ecological and evolutionary qualities of nature:

Our ability to perceive quality in nature begins, as in art, with the pretty. It expands through successive stages of the beautiful to values as yet uncaptured by language. The quality of cranes lies, I think, in this higher gamut, as yet beyond the reach of words.

This much, though, can be said: our appreciation of the crane grows with the slow unraveling of earthly history. His tribe, we now know, stems out of the remote Eocene. The other members of the fauna in which he originated are long since entombed within the hills. When we hear his call we hear no mere bird. We hear the trumpet in the orchestra of evolution. He is the symbol of our untamable past, of that incredible sweep of millennia which underlies and conditions the daily affairs of birds and men.

And so they live and have their being—these cranes—not in the constricted present, but in the wider reaches of evolutionary time. Their annual return is the ticking of the geologic clock. Upon the place of their return they confer a peculiar distinction. Amid the endless mediocrity of the commonplace, a crane marsh holds a paleontological patent of nobility, won in the march of aeons, and revocable only by shotgun. The sadness discernable in some marshes arises, perhaps, from their once having harbored cranes. Now they stand humble, adrift in history.[19]

Leopold was not alone in emphasizing the science of ecology, but his powerful metaphors of nature, poetically expressed, certainly helped usher in an age of ecology. Ecology became a respected profession during the latter half of the twentieth century. Ecologists became nature's doctors, and science gave them a social platform from which to speak for nature. The authority of science positioned ecology as a moral counterweight to the laissez-faire economic rationalism that often justifies social policies. For these reasons, ecology has been called a subversive science.[20]

Ecological thinking raises difficult questions about mechanistic-capitalist assumptions of unending growth and consumption. And some ecologists, as the doctors of Earth, began speaking up for their patient. It is no accident that the Nature Conservancy, one of the largest and more effective non-governmental environmental protection organizations in the world, sprang from the loins of the Ecological Society of America (ESA). The Society for Conservation Biology similarly spilt off in pursuit of a political agenda; it actively promotes a normative science with the goal of conserving biological diversity.[21]

The emergence of groups that intentionally mix policy and ecology represents a tension within the ecological profession. Some ecologists want to keep science neutral and separate from policy. Others, concerned about what they see as an environmental crisis, use the power and privilege of their science to affect social change. In the 1960s and '70s, as the prominence and promise of ecology grew, some ecologists boldly argued that the task of every ecologist is to "inculcate in the government and the people basic ecological attitudes.... Each child should grow up knowing and understanding his place in the environment and the possible consequences of his interaction with it."[22]

Prudence and experience took the edge off some of this obvious advocacy, but the ESA still trains ecologists to work with the media, to become leaders for social change, and to influence policy. A series of articles published in 2000 in the prestigious journal *Science* described some of these programs and ambitions in a section entitled "Ecologists on a Mission to Save the World."[23]

Ecologism

Deep ecology, bioregionalism, and some aspects of sustainable development are political movements based to varying degrees on ecological thinking—referred to here as ecologism.[24] Advocates believe that nature, or ecology as the study of nature, provides a guide for human culture. Ecologism views most other environmental policies as shallow because they merely manipulate environmental systems toward human wants and needs. In contrast, ecologism positions humanity's needs equal to or below the needs of other units of nature and equates ecological health with environmental conditions reflecting little or no human impact, such as those present in 1491.[25]

The social/political implications of ecologism are profound: Humans have no more rights than other units of nature; human population must be reduced to lessen the burdens placed on other units of nature; human economy must be redefined so that it strives for sustainability and cooperation rather than growth and competition; and, finally, functioning ecological systems should replace technologically improved ones.

Earth First!, Earth Liberation Front, and related movements advocate active opposition to the global techno-economic systems that supposedly degrade, denigrate, and destroy Earth, animals, Mother Nature, and other ecological wholes. They argue that the dominant ideals of a consumptive culture exploiting a mechanical nature are fatally flawed. The current system is so badly broken that it must be abandoned. Some of these organizations support acts of physical violence that destroy, sabotage, or otherwise incapacitate industries they consider exploitative. No rebellious act is too small or too big: windows of fast-food restaurants are smashed, farm animals released, biotechnology research labs burned, ski lodges that have been built on habitats of endangered species destroyed, forest industry offices trashed, and commercial fishing vessels sunk.

Defenders of the status quo label these actions ecoterrorism. Others call them eco-revolutionary, fighting against tyranny to save Earth. The websites and publications of these organizations do more than explain how to conduct these terrorist or revolutionary acts. They also describe legal, nonviolent actions people can take to hasten change, such as vegetarianism, organic agriculture, and industry boycotts. The logic and rationale behind these movements focus on the social inequities caused by a globalized market economy, animal pain and suffering caused by agriculture and science, and future generations being shortchanged by current greed and selfishness. Concerns about wild nature, ecological integrity, Mother Nature, the balance of nature, and other forms of ecological holism interweave with concerns about social equity.

Not surprisingly, the policies of ecologism have attracted criticisms. Popular critiques include in their titles phrases such as "Playing God," "The Rising Tyranny of Ecology," "Saving the Environment from the Environmentalists," "Dismantling the Fantasies of Environmental Thinking," and "Green Delusions."[26] These critiques question whether ecological principles should be transferred to social-cultural systems and liken ecological fascism to social Darwinism. *State of Fear* is a recent bestseller by *Jurassic*

Park author Michael Crichton that puts this critique of environmentalism into the public eye. It portrays environmentalism in the same light as religious fanaticism by telling a fictional story of environmental groups so blinded by faith in nature that they intentionally create floods and tsunamis to kill people and attract media attention to what they see as the unforgivable damage that humanity is wreaking on Mother Earth.

CONCLUSION

Ecology and evolution teach us that we live in a dynamic and interconnected system, where change is the norm and surprise is inevitable. Scientific ecological knowledge will be imperfect because uncertainty is part of any dynamic and random system. Thus, it may be prudent to be cautious. Leopold warned that an intelligent tinkerer saves all the pieces. If we disassemble a watch and reengineer it to be more efficient, we should save the parts not used, just in case the new timepiece stops working and we still need to measure time. Thus, we should be cautious when removing "weed" species, replacing natural processes with technological innovations, and otherwise simplifying and modifying functioning ecological systems.

Obviously, we are unlikely to achieve sustainability if we ignore ecological science. However, ecology alone cannot define the conditions of sustainable and thriving communities. We must always remember that the conceptual models we use to interpret nature—such as wholes, machines, and hierarchies—influence the policies and actions we pursue. These three basic conceptual models for understanding ecological interactions are not right or wrong, just incomplete. The outcomes we desire and the management we practice reflect the models of nature we impose on the situation.

* 5 *

(In)finite Nature

Scarcity • Ultimate Resource
Management Implications • Appropriate Technology
Pandora's Box • Taking Responsibility

Imagine a pond with a single water lily that has the ability to reproduce itself daily: one lily on day 1 turns into two lilies, which on day 2 turn into four lilies, which on day 3 turn into eight lilies, and so on. The pond is finite, so exponential population growth can't go on forever without causing problems. On the day the pond becomes half full, one may look around and see plenty of open water and little reason for worry. But by the next morning, the pond's surface will be completely covered and lilies will begin to die because the pond's carrying capacity will have been exceeded (a completely covered water surface chokes off the sun and air on which the pond's ecological systems and lilies depend). Do humans face a similar dilemma?[1]

Obviously the analogy is not perfect. Humans are different than lilies; we use technology to extend and replace finite resources. Moreover, Earth is bigger than a lily pond. Still, difficult questions remain: What are the limits of Earth's carrying capacity, and when should we take serious action to avoid disaster? Human population took several million years to reach 1 billion people, perhaps sometime in the early nineteenth century. It took only 130 years or so to double to 2 billion, sometime around 1930. World population doubled again about forty years later, in 1974. In 1999 we exceeded 6 billion people and are currently adding about 200,000 people each day.

Must humanity limit its growth to avoid running out of resources and exceeding Earth's carrying capacity? Or is the ultimate resource human creativity? Will new technologies provide abundant energy, substitutes for scarce resources, and a clean environment? Answers to these questions will only be known with hindsight, so arguments about these issues are based largely on matters of faith—faith in our technological prowess. This faith provides a critical fault line in most environmental debates. This chapter explores the role of technology in shaping environmental debate by exploring three issues: (1) concern that increasing population pressures and consumer demands will eventually exceed nature's productive capacities and endanger civilized society, (2) belief that human creativity will overcome resource scarcities and pollution problems, and (3) understandings of nature and expectations of management shaped by the role we expect technology to play in creating a sustainable future.

FINITE NATURE

American settlers conquered nature with a vengeance. There was little concern about running out of resources in the face of nature's abundance. Citizens were charged with the "patriotic duty" of conquering nature to fuel the economy and build a civilization. The ingenuity and dedication devoted to these tasks were impressive. By the end of the 1800s, farmers had settled the frontier, transportation networks had united the nation, and industrialization had supercharged the economy (chapter 6 reviews this history). These developments placed voracious demands on nature, raising concerns that America's resources might soon be exhausted. Just as a candle consumes itself, lack of resources could extinguish the hope and promise of America. Observers started worrying about eroding soil, timber famines, and a vanishing frontier. Rather than being infinite, people started talking about nature as being finite, frail, and scarce.

Massive social disruptions resulted as the rural, agricultural economy rapidly transformed into an urban, industrial, supercharged economy. People were displaced, hungry, and destitute. Rural people fled to urban centers seeking employment, food, and shelter. Cities were not prepared to handle the flood of migration and waste. Horrific living conditions demanded attention. Streets were strewn with mounds of rotting garbage and horse manure, while open canals drained raw human sewage. Hair and other nonmarketable animal parts rotted or burned in open pits. Industrial wastes flowed directly into rivers, and acidic smoke and particles polluted the air. Drinking water spread cholera and other toxins, and cramped living conditions spread infection. Factory accidents crippled, industrial chemicals poisoned, and scarce open space confined a weary and impoverished working class. Social services were abysmal, with inadequate schools, high crime, poorly enforced sanitary regulations, little lighting, unpaved streets, political corruption, and unchecked epidemics.[2]

These environmental degradations motivated pessimistic predictions that human-population pressures would eventually exceed Earth's carrying capacity. Similar claims have been made throughout history, perhaps most famously by the Reverend Thomas Malthus of England, who in 1798 published an essay arguing that nature did not have an infinite ability to support human population growth. Reverend Malthus argued that the Creator, rather than being benign, crafted a system whereby populations naturally grew at rates that eventually and inevitably exceeded the capacity of their habitat to provide space and food. He wrote that these natural "laws" pervade "all animated nature." Rather than instilling balance, the Creator gave all species "a power of a superior order" to reproduce in excess of a habitat's capacity to produce food.[3] The inevitable result of the imbalance is starvation, strife, and death. His studies of nature found numerous examples of these boom-and-bust cycles. Animal populations explode in times of plenty and collapse after resources have been exploited. Malthus asked not whether this would happen to humans, but when. His dire forecast of inevitable resource scarcity and human suffering earned him some fame because it ran counter to the prevailing theme of prosperity and growth promised by modernity. He revised and republished his arguments several times and now similar arguments are frequently dubbed "Malthusian."

In 1864 George Perkins Marsh published one of the most important and influential environmental books, *Man and Nature*. Amazingly, it became a bestseller, even though at the time the public was distracted by the Civil War and aggressively committed to industrialization. Marsh warned Americans of their precarious dependence on healthy, functioning, and productive environments. As U.S. ambassador in Turkey and Italy, he observed severely degraded environments caused by generations of exploitation. He hoped to prevent the United States from experiencing a fate similar to the Roman Empire, whose fall he attributed to poor stewardship of environmental resources:

Vast forests have disappeared from mountain spurs and ridges; the vegetable earth accumulated beneath the trees by the decay of leaves and fallen trunks, the soil of the alpine pastures which skirted and indented the woods, and the mould of the upland fields, are washed away; meadows, once fertilized by irrigation, are waste and unproductive, because the cisterns and reservoirs that supplied the ancient canals are broken, or the springs that fed them dried up; rivers famous in history and song have shrunk to humble brooklets; the willows that ornamented and protected the banks of lesser watercourses are gone, and the rivulets have ceased to exist as perennial currents, because the little water that finds its way into their old channels is evaporated by the droughts of summer, or absorbed by the parched earth, before it reaches the lowlands; the beds of the brooks have widened into broad expanses of pebbles and gravel, over which, though in the hot season passed dryshod, in winter sealike torrents thunder, the entrances of navigable streams are obstructed by sandbars, and harbors, once marts of extensive commerce, are shoaled by the deposits of the rivers at whose mouths they lie; the elevation of the beds of estuaries, and the consequently diminished velocity of the streams which flow into them, have converted thousands of leagues of shallow sea and fertile lowland into unproductive and miasmic morasses.[4]

People had witnessed in their lifetimes environmental changes of shocking magnitude. The rapid decline and extinction of the superabundant passenger pigeon is just one example. Early observers described flocks that turned the sky black, were as loud as tornados, broke tree branches under their weight, and caused horses to stop and tremble. One flock was estimated to be 240 miles long, containing over 2 billion birds, and taking over five hours to pass. Conservation efforts seemed ridiculous with no end of these flocks in sight. Pigeons were shot for target practice, and people feasted on pigeon pie. Trees where flocks roosted were set ablaze to cook and kill thousands more birds than could be eaten or processed, and farmers fattened their pigs on the dead birds.

The pigeon population declined rapidly, from billions in the 1870s to dozens in the 1890s. The last passenger pigeon died on September 1, 1914, in the Cincinnati Zoo. Other species fell into or came dangerously close to oblivion, and the previously unknown concept of extinction emerged as a topic of concern.[5] The once-abundant nature was depleted within an eye blink of history. Citizen groups formed in response to cries of alarm. One of the first environmental groups, the Boone and Crockett Club, focused most of its attention on declining wildlife populations.

Our technology seemed unable to solve these grave environmental problems. In fact, in many cases it caused more harm than good. In May 1934, 350 million tons of the nutrient-rich, crop-producing, economy-sustaining, nation-feeding, midwestern soil blew east. Twelve million tons of it fell on Chicago. Dust clouds darkened skies over Boston, New York, and Washington, D.C., and ships in the Atlantic reported dust on their decks.[6] How could scientific agriculture produce such catastrophe? When Lewis and Clark explored the Pacific Northwest in the early 1800s, perhaps 16 million salmon annually spawned in the Columbia-Snake river system. Regional cultures and economies grew up around these rivers and their fisheries. How did it come to be that in 1994 observers standing at the first dam on the Snake River spotted only fourteen salmon? Dozens of agencies, thousands of environmental professionals, and billions of dollars had not stemmed the decline.[7] If our best science and technology were unable to prevent these high-profile catastrophes, then what else might be going wrong?

Alarmists say the warning signs are growing more ominous. A thinning ozone layer allows life-threatening radiation through to Earth's surface, and increasing carbon dioxide fuels global climate change. A two-mile-thick brown cloud of industrial pollutants and ash floats above the entire

Indian subcontinent, blocking out 10 to 15 percent of the sunlight, shrinking agricultural productivity, and causing erratic weather patterns. Similarly, an oxygen-starved "dead zone" the size of New Jersey appears most summers in the Gulf of Mexico, causing massive deaths of aquatic life. Rain, acidified by fossil-fuel emissions, destroys life in remote lakes and forests. Perhaps our inability to find technological solutions to these pressing environmental ills provides ample reason to worry about hubris. Did Pogo get it right when he observed: We have met the enemy and it is us?

Malthusian-type concerns about human population growth re-erupted in the 1960s. Popular publications argued that population "bombs" and "explosions" would ultimately cause environmental disaster, panic, and give rise to "lifeboat ethics." Supposedly, we would fight wars to secure food and other scarce resources, force sterilizations and abortions to reduce population growth, and turn a blind eye to malnutrition and dehydration in developing nations. The space age provided a sobering image of a finite, lonely, beautiful blue-green orb floating in the endless black of a hostile space. It also provided another metaphor: "Spaceship Earth." A spaceship is a closed system because it carries with it all its resources and recycles many of its wastes. A spaceship's economy must minimize energy and material throughput and maximize the stocks of finite resources. This space-age image and metaphor challenged the dominant "cowboy economy," which assumed an endless supply of energy that could support an ever-expanding population and economy.[8]

In 1972 a group of computer modelers published a study titled *Limits to Growth*. Their models simulated the interactions between population, agriculture, natural resources, industry, and pollution and confirmed Malthusian concerns by showing how population growth would outstrip food and resource production. They concluded that we had less than a century if current trends went unchecked, otherwise scarcities and pollution would cause serious declines in quality of life. The future looked gloomy even if technology could double the amount of each resource (i.e., more fish from the sea, bigger trees from the forest, fresh water from oceans) because the amount available to each person would decline with so many more people to feed, clothe, and house.[9] The message is the same, regardless of whether the point is made with a lily pond, a spaceship, or a computer model: Exponential growth cannot continue indefinitely on a finite world.

NATURE MADE INFINITE THROUGH
TECHNOLOGICAL INNOVATION

Few people dispute that Earth resources are finite; pictures from space are dramatic reminders that we have but one planet receiving a finite amount of solar energy. So the real arguments are not about whether there exist infinite resources, but how well human technologies or supernatural interventions can extend, conserve, and find substitutes for these finite resources without sacrificing human quality of life. Faith that God will provide is discussed in chapter 9; faith that human technology will provide is the topic of this section.

Techno-optimists argue that technology creates new resources to keep pace with population growth. They believe human creativity is the ultimate resource, not soil, water, solar energy, or fossil fuels. Necessity is the mother of invention: if and when resource shortages or pollution hazards become acute, solutions will be generated by creative minds motivated by the great wealth that the market economy bestows to people who meet important human needs. We do not need to conserve resources for the future because people of the future won't need those resources. Substitutes will be found for anything that becomes scarce. The current abundance of fossil fuels, for example, provides the opportunity to build research labs, complex machinery, and the educated workforce needed to harness more sustainable and less polluting energy sources such as fusion, gravity, wind, and solar radiation. In the minds of techno-optimists, these infinite and very clean energy sources combine with advances in bioengineering to promise a nearly unlimited supply of food and building materials. The sky is the limit, and perhaps not even that. The ultimate resource is human creativity.

Techno-optimists also believe technological advances will clean and improve environmental quality. For example, the energy efficiency of lights and appliances has increased dramatically in recent decades. Bacteria and other microbes detoxify pollutants and create useful products from what was previously discarded as wastes. Green chemistry makes new chemicals safe and plastics biodegradable. Smokestack scrubbers clean air emissions, and sewage is processed into drinking water. Nanotechnology builds ever more complex structures using significantly fewer resources, while new factories replace older, inefficient, and polluting technologies.

The promise of technology is that it can solve seemingly insurmountable environmental problems such as global deforestation and biodiversity loss. For example, estimates suggest that we could meet the worldwide demand for wood by intensively managing as tree plantations just 5 percent of the world's total forestland. Likewise, pesticides, fertilizer, irrigation, and biotechnology allow more and more food to be grown on less and less land. These forestry and agricultural practices free up land for wildlife habitat, biodiversity, water, amenity production, or other concerns.[10]

Few techno-optimists dispute that previous technologies have degraded environmental quality or that eroded soils, toxic wastes, and ozone holes present serious dangers. But the optimists argue that technological advances will improve the environment if and when these problems become socially relevant. After all, it is technology such as satellites and computers that allow us to identify and monitor environmental concerns such as global climate change and deforestation. These technologies give us advantages over previous civilizations that went extinct because they did not know the environmental consequences of their actions. Our technology directs us toward behaviors that enable sustainability.[11]

A growing literature supports assertions that resource scarcity can be solved through technological innovation. It pokes fun at naive assumptions and disproved predictions made by early alarmists, such as those who concluded that America could not sustain a population greater than several million. This nineteenth-century prediction assumed that the limiting factor to sustainability would be the pasture required to feed horses and other beasts of burden, which were then the primary engines behind transportation, agriculture, and industry. Consistent with the techno-optimist position, this limitation vanished with invention of steam and internal combustion engines.

The Malthusian prediction that resources would grow scarce and more expensive with population increases also has not proved true. Several studies have shown that the amount of labor and investment capital needed to grow crops, mine materials, and transport those resources to consumers have actually decreased. For example, from the time of the Civil War until now, the availability of grain, sugar, cotton, meat, dairy, copper, petroleum, coal, and lumber has increased while the relative price has decreased. Studies also show that during the latter half of the twentieth century, the environment got cleaner and healthier rather than more polluted and degraded. The explanation for these anti-Malthusian findings is that technology increases resource yields and reduces pollution faster than population increases resource demands and environmental degradations.[12]

IMPLICATIONS FOR UNDERSTANDING NATURE

Being an optimist or a skeptic is a matter of faith. It is a belief based on assumption and intuition, and there is no way to prove whether the faith is justified until we have the benefit of hindsight. It is clear, however, that faith in technology plays a major role in understanding nature and defending positions in environmental debate, which are the topics of this section.

Before proceeding much further, it seems appropriate to define technology. Technology, broadly defined, is the product of human creativity. People invent and apply technologies to manipulate natural and social systems to satisfy some goal: convenience, wealth, safety, entertainment, or the like. Consider the classic dilemma of overgrazing the community pasture with too many cows.[13] Visualize an idyllic small town where residents graze their cows on a common pasture. Assume the pasture is currently grazed at capacity: each cow gets all the grass it needs, no excess grass is left, and any more grazing would degrade the pasture's ability to sustain current levels of grass growth. The tragedy of the commons occurs because town residents, acting rationally for their own self-interests, add more cows than the pasture can sustain. These actions are rational (for the individual in the short term) because each additional cow benefits its owner more than it costs its owner. The benefits of the extra cow go only to its owner, while the costs of degraded pasture and slightly lower milk productivity are shared by everyone. If many residents follow suit and add more cows, the pasture will collapse from overgrazing, all the cows will starve, and everyone will lose.

How might technology solve this problem? Direct manipulation of physical and natural systems using engineering and natural sciences provides one solution. For example, engineering technologies such as irrigation and fertilization might create more productive pastures, or genetic modifications might make cows more efficient at converting grass into protein. These advances might only postpone the tragedy, however, because residents can continue adding additional cows to the commons until the pasture's capacity is again exceeded.

Other solutions to this problem include laws, regulations, and education. These social technologies are among our most powerful problem-solving tools. While engineers may point to the wheel, printing press, computer, antibiotics, and the combustion engine as examples of technologies that changed the world, the impacts of these inventions pale in comparison to language, law, democracy, capitalism, art, and science. Social technologies such as environmental education teach cow owners about the potential tragedy of overgrazing the commons, tax incentives and coercive regulations encourage fewer cows, vegetarian diets reduce the demand for cows, and property privatization gives each family control over a portion of the pasture and creates selfish reasons for conservation. We have a tradition of debating applications of our social technologies (i.e., taxation, environmental regulation, population control, etc.) more publicly than we debate application of engineering solutions, but that may be changing with increased public concern about genetic engineering, pesticide approval, and food labeling.

Faith in Technology Influences Policy and Management Options

Defining sustainability is one of the most pressing and perplexing tasks facing society. Different expectations about technology produce very different definitions and hopes for sustainability. One of the common definitions of sustainability is "Sustainable development is development that meets the needs of the present without compromising the ability of future generations to meet their own needs."[14] This definition allows technological advances to substitute for degraded environmental qualities. Rather than passing on to future generations functioning ecological systems that meet their needs, we can instead pass onto them the technological capacity to reengineer systems so they can meet needs as they define them. For example, techno-optimists argue that technological advances will create clean and abundant sources of energy that will replace fossil fuels, mitigate global warming, and cure social ills. Cheap and clean energy will power water desalination and purification, thereby resolving many of the most pressing social and environmental threats to human health. Genetic engineering will create bacteria that consume toxins and superproductive plants that grow all the food we need on small areas of cultivated land. Techno-optimists therefore want to provide future people with increased technological capacity to meet future unknown needs. Techno-skeptics don't trust technological solutions and instead want to sustain the ecological systems that have historically provided clean water, climate stability, and soil production. Techno-skeptics don't want to force future generations to be reliant on energy-intensive technological solutions for their survival, especially because such systems can be oppressive, rigid, and brittle.

Techno-skeptics worry that technology creates unacceptable risks. We have only one planet, so we ultimately have but one experiment. They advocate applying a precautionary principle: new technologies or environmental changes must be proven safe before wide distribution. Techno-optimists counter that the precautionary principle slows progress and application of technology that may improve quality of life. They don't want to delay opportunities and rewards that might be forthcoming by advancing technology and managing nature. They point out that the precautionary principle is nothing more than a value judgment about risk. Techno-optimists believe that the risks of forgoing new technologies outweigh the risks of environmental damage.[15]

Techno-skeptics argue that technological solutions often create more problems than they solve. As a result, they prefer cautious observance of traditional land-use practices that have worked in the past. In other words, they would rather mimic a tried-and-proven nature than reengineer it. They are more likely to support preservation than development, and when development is necessary, to prefer small-scale projects that don't rely on complex, interconnected technological systems that could fail catastrophically. Techno-optimists, in contrast, have confidence that human technology will solve problems if and when the problems become critical. It is inefficient to worry about or direct resources toward solving environmental problems that are not yet critical. The market will reward the solver of critical problems with great riches, motivating society's best and the brightest to focus attention on socially relevant issues. As a result, techno-optimists are more likely to support land-development projects that involve large-scale changes and projects that reengineer natural systems to make them more efficient and productive.

Techno-optimists are more willing to rush toward a technological utopia. Assuming that production efficiencies and product substitutions will alleviate resource scarcities, they worry less about preserving natural systems that produce valued ecosystem services and worry more about delaying realization of the untapped potentialities of bioengineered nature. Techno-optimists want to fix the inefficiencies of nature to increase resource yields, whereas techno-skeptics worry about getting trapped in an endless cycle of needing new technological advances to solve problems created by past technological advances. More technology requires more complex social structures, each added complexity producing less return. Gains from the low-hanging fruit of fossil fuels and computer technology have been captured. Comparable technological advances will require massive investments of social capital and thus may cost more than they return. For example, the gains from new technological advances in agriculture, medicine, and science have steadily declined over the last hundred years or so, while costs required to produce each new advance have increased.[16] The increasing dependence on technology creates an increasingly complex and brittle system that is vulnerable to collapse. Rather than pursuing new technological advance, skeptics argue that we need to slow down or change direction.

Techno-optimists believe that technology increases opportunities and expands horizons. It frees us from dangerous and dulling labor and inspires new dreams about human potential. Technology creates the free time that allows us to become educated, practice the arts, and dream of utopia. Techno-skeptics, on the other hand, argue that technology limits options by restricting our view to solutions enabled by the technology we've mastered. Carpentry hand tools provide a common illustration. If your only tool is a hammer, everything becomes a nail. Possible designs for furniture or structures are restricted to the hammer technology of attaching items with friction and wedges. Screws, bolts, pulleys, glues, and other workshop technologies are useless and ignored until the craftsperson learns to use drivers, wrenches, ropes, and clamps.

Numerous examples of how technology restricts our solution set can be seen in environmental management. If we know how to reduce insect-caused crop losses by applying chemical pesticides, these chemicals become our hammer, so we continue to invent and apply more of them. Integrated pest-management strategies may be cheaper and more effective but are ignored because they don't fit into the normal way of thinking about the problem. If we know how to limit pollution by regulating industrial discharges, government regulation becomes our hammer, so we place more limits and heavier fines on polluters. Proposals to redesign production systems are shelved, even though they minimize or recycle material consumption, thereby reducing wastes and saving money. Successful, established technologies become the tools in the professional's tool bag. When asked to solve a problem, the professional applies tools that worked in the past and it becomes difficult to think outside the box. Innovation is risky.

Appropriate Technology

Of course, the real world is not nearly as black-and-white as presented here. Nature is not infinite *or* finite and technology is not good *or* bad. These are false dichotomies. Instead, there exist various shades of gray, or in this case green. Hard green and soft green illustrate competing paradigms for thinking about the role of technology in solving social and environmental problems. Both approaches advocate technological solutions, but the solutions differ dramatically, as do their implications for society. Soft-green technologies attempt to mimic nature and minimize disruption to natural systems. Hard greens want to learn from natural history but not be limited by it. Soft greens address finite energy concerns by decreasing energy consumption with technologies such as passive solar heating, while hard greens seek to increase the energy supply with clean technologies that use the enormous coal reserves or that safely tap nuclear energy. Soft greens maximize recycling and eliminate waste; hard greens bury wastes that can't be efficiently used in today's economy with today's technology. Problems caused by these wastes will be solved as they arise, and the wastes might even be mined as new technologies and markets make them valuable.[17]

Bio-mimicry is a soft-green technology that attempts to mimic and work with natural systems rather than simplify functioning biological systems with machinelike technologies that depend upon cheap inputs of fossil fuels and ignore the consequences of polluting outputs. It recognizes that functioning biological systems, refined over time by countless tests of natural selection, have developed effective means to generate and store energy, catalyze chemical reactions, store and transport water, recycle nutrients, resist infection, and perform countless other tasks essential to sustaining humanity. Bio-mimicry seeks to partner with systems designed by evolution.[18]

"Cradle to cradle"is a philosophy that allows consumerism to continue but seeks to *eliminate* rather than merely reduce wastes from the manufacturing process. It combines both hard- and soft-green technologies. Products and by-products of manufacturing processes are thoughtfully designed so that they feed into other production systems. Wastes become the nutrients of new products. Recycling, rather than "down-cycling," is the goal. Down-cycling occurs when materials are mixed or contaminated during manufacturing and cannot be returned to primary products but instead are turned into lower-grade materials that eventually head for the dump. For example, high-grade plastics get down-cycled into plastic bags or packaging peanuts. If recycled again, these down-cycled plastics get further down-cycled into extruded wood that has no recycle potential. At each cycle the materials get more diluted or contaminated and less useful to manufacturing. Advocates of this philosophy see a rosy future where appropriate technology improves both environmental and human conditions:

> We see a world of abundance, not limits. In the midst of a great deal of talk about reducing human ecological footprint, we offer a different vision. What if humans designed products and systems that celebrate an abundance of human creativity, culture, and productivity? That are so intelligent and safe, our species leaves an ecological footprint to delight in, not lament?[19]

Advances in manufacturing technology that eliminate waste are necessary but insufficient means to create a cradle-to-cradle consumer culture. A cultural revolution is required to change the way we buy and sell products. Manufacturers and consumers would need to be convinced to sell and buy service contracts instead of products. In a cradle-to-cradle economy, consumers upgrade their service contracts to obtain the latest and greatest styles, performance, and conveniences. They do not discard old products and purchase new ones. For example, you could rent a TV for 10,000 hours or a car for 100,000 miles and return it to the manufacturer when you renew the service contract. Your new contract would provide a refurbished appliance with the latest and greatest features. Consumers wanting new features before their contract expires simply upgrade their contract, for a price, and receive a replacement appliance with the latest features. The manufacturer, rather than constantly seeking new resource streams to make new products, would design and use the parts of the old appliances as "nutrients" for next-generation products. Such a philosophy obvious overlaps with green consumerism (reviewed in chapter 6), which suggests sustainable consumption is not about consuming less; it is about consuming differently.

The contrasts between hard- and soft-green technologies are not trivial. Promoting one approach over the other will produce very different futures, with different risks, different environmental qualities, and different opportunities. Society's task is to define a role for the machine in the garden, such that the machine does not destroy the garden and perhaps even improves it. Our search is for technology that is appropriate to the social and environmental goals we hold.

Some technologies, because they are enormously complex and expensive, require centralized authority and extreme expertise. Becoming reliant on these technologies may create impersonal and oppressive social systems, restrict personal freedoms, and snuff out local diversity. For this reason, social critics warn us to explicitly consider the social implications of our technological choices. "Small is beautiful" and "act locally" are their watchwords.[20] They advocate technologies that, in addition to being ecologically benign, help communities become self-sufficient and politically autonomous. They advocate technologies that promote diversity, minimize oppression, require low-capital investment per unit of output produced, can be maintained with a nonhierarchical organizational structure rather than a pyramid of expertise and regulatory controls, and create local employment opportunities. Solar cells and wind farms are examples of power sources that can be decentralized and locally maintained, whereas nuclear power requires reliance on a large power distribution system and decision makers living outside the community. Technology thus affects not just how we manage the environment but also how we organize ourselves and our future opportunities.

Defining Nature

Society directs wealth and status toward those who solve threatening problems such as timber famines, water pollution, energy crises, and food shortages. Numerous environmental sciences and professions exist as a result. Professionals now provide environmental organizations, companies, and land-management agencies with expertise such as urban planning, public health, civil engineering, hydrology, forestry, and agronomy. These professions define nature as something that can be sustained and protected through careful professional management and define environmental degradation as pollution that can be controlled and cleaned up or inefficiencies that can be reengineered. Their social status makes them not just the protectors and managers of nature, but also gives them the power to define nature using their scientific and technical language.[21]

Nature is known to most environmental professionals through the lens of applied science and technology. Environmental science tends to focus on observable, measurable phenomena, such as the number of trees, the depth of soil, the variety of species, the amount of water, the nutrients that are exchanged, and the genes that organize organisms. Sciences that serve economic interests, such as forestry and agriculture, further restrict their focus to entities and processes of nature that have economic value, such as board feet of timber, bushels of corn, and site productivity.

There is nothing wrong with these scientific and professional descriptions of nature, but they are partial. People with other agendas and worldviews might value and want to talk about different aspects of nature not easily captured by these applied, problem-solving sciences. For example, nature known from the perspective of equity might focus on quality of life and justice resulting from environmental change; nature known from the perspective of spirituality might focus on feelings of inspiration and communion with a larger whole; similarly, nature known from the perspective of aesthetics might focus on feelings of beauty, solitude, and relaxation, while nature known from the perspective of identity might focus on the memories of people and events attached to a place. Subsequent chapters explore these and other understandings of nature. The dominance of the techno-scientific language, as well as the high status of the natural sciences, makes it difficult for these other understandings of nature and their advocates to compete for attention in environmental planning efforts.

To better understand the ability of technology to restrict our attention to particular slices of nature, consider the following example. Environmental science prior to the 1960s developed concepts and understandings of nature that served the political agenda of progressive conservation, which focused on minimizing waste and maximizing efficiency. Science developed constructs, theories, and methods that concentrated on the sustained yield of socially valued environmental qualities (so-called resources such as air, water, wood, and wildlife). As a result, we had lots of information about natural resources and the factors affecting their yields. But during and after the 1960s, ecological understandings of nature motivated scientists to appreciate different aspects of nature, such as nutrient cycles, genetic drifts, carrying capacity, and biodiversity. The resulting ecological science produced a new language with which to describe nature. This new technology created ways to evaluate environmental quality and conceptualize environmental policies that preserve biodiversity, protect wetland functioning, and maintain ecological integrity. Different science, motivated by different reasons, emphasizes different understandings and descriptions of nature.[22]

Anti-technology arguments point to humanity's immaturity and lack of restraint. Technology may give us the power to change the world, but some people worry that we do not possess the maturity to use that power wisely and instead might open Pandora's box and release the agents of our demise. Exploding atomic energy in an act of violence, however necessary to end World War II, was followed by an end-of-the-world defense policy of mutually assured destruction, suggesting that perhaps we are not yet ready to deal with the powers some technologies offer.

The techno-skeptics argue that caution and prudence are needed, not reckless advance. Techno-optimists, in contrast, contend that we have rarely if ever had the ethical foundation to evaluate the changes we make to nature until well after the changes have become mainstream. Had we waited for ethical debate and erred on the side of caution, we might still be short-lived, illiterate cave dwellers. Our ancestors no doubt experimented with fire, tools, language, and agriculture without much concern for the implications of these biosphere-changing technologies. Because they took risks and made innovations, we now have the leisure, means, and need to debate environmental ethics. Our urge to invent, experiment, and strive for something better is what created us—the tool-using, big-brained, ecosystem-altering human species. Our technologies shaped our evolution and our ecology. We must now learn how to live with technologies that have the potential to end our evolution and our ecology.

Taking Responsibility by Controlling Risk

Nature's dynamics create risk: risk of famine, disease, flood, and pestilence. Controlling nature's dynamics can minimize risk. Agriculture, for example, increases our control over the uncertainties of finding sufficient wild foods; medicine increases our control over disease; dams increase our control over flooding; and automobile and information highways increase our control over space and time. With this control comes responsibility over matters of life, death, and future generations. A medical example illustrates the challenges, responsibilities, and moral dilemmas that technological advances can create.

Cystic fibrosis is a genetically transmitted disease carried by people of European descent. It causes problems with digestive and respiratory systems that lead to infection, malnutrition, and death. Before modern medicine understood the disease, afflicted babies died within their first years of life. Knowledge of the disease now provides some measure of control that forces difficult choices.[23]

Scientists found a genetic indicator of the disease in the 1960s. This awareness presented potential parents carrying the gene with the difficult choice of risking having a child with the disease, being childless, or choosing adoption. Not all offspring inherit the problematic gene and improved technology enabled testing fetuses in the womb, presenting parents with another difficult opportunity: conceive, test, and abort afflicted children before birth. More recent technology allows in vitro fertilization, providing parents with the opportunity to conceive a child using someone else's genetic material. Future developments in genetic engineering may give parents the opportunity to remove just the disease-causing genes, raising troubling questions about parentage and eugenics. Medical advances in care for the disease further increases the dilemma. Life expectancy of the afflicted is now thirty or more years with greatly improved quality of life. Now parents must factor into their decisions the extent of pain, suffering, love, and life they want to share or deny. Each advance of technology provides more control over nature and greater awareness of risk. Control and awareness create opportunities, which in turn require choices. Situations once left in the hands of chance, fate, and God are now under our control.

Environmental professionals are faced with similar ethical dilemmas. The chemical revolution resulted in widespread application of pesticides and other chemicals that posed countless health risks. However, the increased agricultural productivity enabled by these chemicals might have saved human lives with improved nutrition and saved millions of acres of wildlife habitat that otherwise would have been cleared. Transgenetic engineering also presents dilemmas. The world population continues to increase and food production must keep pace. Some engineered plants dramatically reduce the need for pesticides, tilling, and other potentially damaging practices. Further advances in genetic technology could feed the world on fewer and fewer acres, increasing the habitat for wildlife and opportunity for spiritual, aesthetic, and moral experiences derived in wild places. Yet these advances have associated risks. Understanding and evaluating these risks are key challenges for environmental management.

CONCLUSION

Americans have enjoyed unprecedented affluence and stability enabled by control over economic, social, medical, industrial, agricultural, and environmental systems. But the technology providing this control might be a two-edged sword. Engineered viruses, cloned animals, invaded privacy, addictive entertainment, information overload, and increasing complexity raise concerns about hubris. Exploding atomic bombs, opening ozone holes, and extinguishing species remove any doubt that our cleverness can destroy life on Earth. Is the Enlightenment dream of using science and technology to create heaven on Earth turning into a ghoulish, suicidal nightmare? Humanity faces the fundamental challenge of the child in the candy store. We must restrain our urges to taste it all because doing so will likely make us sick and get us kicked out of the store. We must develop the patience and restraint needed to survive the powers and potentialities that technology promises.

Popular concerns about Frankenstein technology parallel the rising concerns of many scientists about their sciences. As scientists level their critical and skeptical eyes at their own fields, they increasingly recognize the extent that uncertainty limits the role science can play in environmental decisions.[24] Uncertainty in the face of overwhelming complexity is increasingly accepted as a given. Nature is so utterly complex, interconnected, chaotic, and dynamic, that relative to what may be known about it, we now know very little, and we are not likely to ever know all that much.

Science and technology provide powerful solutions to many problems and create many opportunities for improved quality of life, yet they cannot answer the big questions about which future we should strive to create. One way to study these bigger questions is to simulate and forecast possible futures. The predictions will most likely be wrong because of the complexity and dynamism of humanized ecological systems, but they still force us to make explicit choices about which futures we want to create and to measure our progress toward those futures. Improvements in science and technology will make these predictions more accurate, while improvements in self-understanding and public debate will refine our abilities to articulate desired future conditions. Such an incremental, adaptive process helps us learn about the qualities we value and want to sustain as well as how to manage humanized ecological systems in ways that sustain these qualities.

Does the unknowable complexity and unpredictability of nature demand caution and prudence? Probably. Hubris can be dangerous, and some degree of caution seems reasonable. But how much caution? How much should we slow science and technology with the burden of increased regulation and prudence? These are difficult questions to answer because the costs of being wrong are profound. Arrested technological advancement can impose significant hardships. Medical advances might be delayed and unnecessary deaths result. Agriculture crops viable in regions suffering famine might go undeveloped. Unnecessary worker injury and monotony might be suffered because advanced technologies are not integrated into the workplace. New opportunities of education and entertainment might go unrealized for fear of poisoning our minds. However, questioning technological advances also seems wise because the cost of environmental collapse is unacceptably high. We have but one planet with which to experiment. Few issues are more polarizing. And few deserve more discussion.[25]

* 6 *
Economic Nature

Manifest Destiny • Livestock
Sustainable Development
Nature's Services • Greening Capitalism

Capitalism is perhaps the most powerful organizing force in today's world. It transcends environmental and political boundaries by organizing and directing the flow of resources, people, and information. It provides a common language and is more pervasive and influential than any previous social system, even those imposed by Islam, Christianity, Greece, Rome, or Britain. If the world faces environmental problems, then their cause and solution involve capitalism.

This chapter explores how capitalism shapes the way we know and use nature. First it reviews how European settlers tamed and harvested a seemingly super-abundant nature in order to build an economy and establish a culture. Second, it looks at how economics transforms and narrows our understandings of nature: forests become woods, wildlife become livestock, nature becomes resources, ecosystem processes become economic services, and humans become labor. Third, it examines the idea of sustainable development. Fourth, it considers the work of environmentally motivated economists attempting to direct economic markets toward sustainable ends by pricing the role that ecological services play in sustaining human economy and habitat. And finally, it explores green business products and practices attempting to merge profit motives with sustainability concerns.

MANIFEST DESTINY

From our twenty-first-century perspective, it is perhaps impossible to comprehend the challenges and difficulties that nature presented to European settlers. We now have an activist government that provides an (admittedly imperfect) safety net; established utilities that provide transportation, water, sanitation, and power; and an agricultural industry that provides ample food. Except for natural disasters, we don't think of nature as harsh, indifferent to human existence, and the cause of much suffering. But recall that many early European settlers died and that many who did not die owed their survival not to the bounty of nature but to the food they borrowed or stole from Native Americans. In 1587 all 117 settlers of the "lost colony" in North Carolina apparently perished. The first "successful" settlement of Jamestown, Virginia, fared slightly better: only 66 of the original 104 members died in the first year. Early explorers such as de Soto died of exhaustion and exposure. Nature was something to be conquered and converted into civilization-building resources. Kill it or it killed you; tame it or tame your aspirations. Nature was not something with which you identified or empathized. It was not something to which you belonged. It was the enemy in the battle for survival and the raw material for building a civilization.[1]

As Europeans settled the Americas, trees, soil, water, and wildlife seemed stockpiled, abundant, and aching to fuel an economy—running out of them was unimaginable and to waste them was virtuous. For example, trees too large to convert into heat or lumber with ax or saw were a nuisance. They were killed by stripping away a ring of bark completely around the tree, a practice called "girdling," which disrupts the flow of water and nutrients between roots and leaves. The desired effect was a dead, leafless canopy allowing sunlight to penetrate, converting the dormant potential stored in soil into economically profitable corn. If crop productivity faltered after a few years of erosion and nutrient exhaustion, the settlers simply moved farther west, in search of more fertile soils. In this fashion, the frontier lurched westward, converting one of the greatest deciduous forests on Earth into food and profit.

Cleared fields and tree stumps stood as proud symbols of an advancing civilization. Clearing the land of trees, turning the soil with a plow, building roads, erecting towns, and establishing industry were among the noblest things Americans could do. Untamed nature was an obstacle to be overcome. Economic development was celebrated. The nation's vibrancy grew with each forest converted to agricultural field and each field converted to industry. Wilderness was an enemy of civilization to be vanquished, tree by tree and acre by acre. Pioneers were part of an army defeating nature to build an American paradise.[2] Settlers marched across the continent making the useless useful: converting idle, wasted nature into profit-making, economy-building natural resources. This army brought resources to factories and merchandise to market. It built cities, spawned industries, connected the country with trains and telegraph, and laid the foundation for the Industrial Revolution and an industrial nation of unparalleled scope and power. Developers, hunters, miners, farmers, and foresters all did worthy work converting the fallow factory of nature to the useful, noble enterprise of humanity.

Adam Smith's *Wealth of Nations*, published in 1776, set the stage. His synthesis of Enlightenment thinking and the Industrial Revolution defined progress as a scientific law and unquestioned good. He assumed that economic wealth produced human well-being and that more possessions equaled more happiness. The invisible hand of selfish interests would guide social development because individuals pursuing their interests by accumulating capital and consuming products would stimulate the economy and pull everyone out of poverty. Wild nature provided the raw materials to build a capitalist paradise.

The scope and scale of America's natural wealth exceeded comprehension, motivating empire builders to envision birth of the world's greatest civilization.[3] A nearly unbroken canopy of forest stretched from the Atlantic Ocean to the Mississippi River and was followed by a vast expanse of deep soil, endless prairie grasses, millions of buffalo, and large navigable rivers. The majestic Rocky Mountains with their snowcapped peaks hindered travel westward but provided a steady supply of water for settlement and irrigation. Beyond the Rockies lay more huge trees and tillable soil. At the western edge of the continent, the Pacific Ocean provided shipping access to the Far East. The ability to trade with both the Eastern empires of Asia and the Western empires of Europe promised both riches and political influence.

The vision of a connected continent motivated the fledgling colonies on the Atlantic seaboard to expand westward and gain control over all the resources and riches between them and the Pacific. The great Mississippi basin was the supposed key that would unlock the continent's potential and help settlers build an empire. It was the "valley" spanning the Appalachians and the Rockies, and hence provided trade routes connecting the Atlantic and Pacific oceans. In the visionaries' eye, the Missouri, Mississippi, Ohio, and other interconnecting rivers provided the transportation infrastructure bridging this valley and thereby providing a powerful image of one vast economic body held together by the veins and arteries of transportation. The potential for economic and empire development seemed endless.

A connected continent, a thriving economy, and a united nation became accepted as America's Manifest Destiny. The Lewis and Clark expedition is a famous example of publicly funded efforts to explore the continent for riches and for a means to transport them east and west, north and south. Other public and private efforts aggressively developed water, rail, road, mail, and telegraph connections. Canals and ports were built, rivers and bays were dredged, roads connected farms to markets, and railroads bridged the coasts. Private interests such as railroad companies received landholdings as incentives to connect markets and make possible a commerce-based economy. Agriculture and home ownership were motivated by tax incentives and land-distribution schemes. The Preemption Act of 1841 sold land for $1.25 per acre, while the Homestead Act of 1862 gave it away to anyone who would live on and farm it.

Manifest Destiny even played a role in the rationale behind the Civil War.[4] The immense economic and political potential of the midwestern plains inspired Abraham Lincoln and Stephen Douglas to focus on Manifest Destiny during the 1860 presidential campaign. Secession could not be tolerated. A divided country would fragment the economy and destroy the country's potential. Power, wealth, and empire were possible if the continent's vast interior resources could be used to fuel commerce and spread American culture from the Atlantic to the Pacific coasts. The potential of a thriving economy with which to build a strong nation would not be realized if states controlling the corridors of trade—the Mississippi and other midwestern rivers—could not work cooperatively with the East Coast ports to maximize the economic potential of natural resources, agricultural produce, and manufactured goods. Such economic collaboration seemed less likely and less profitable if the country were divided and trade subject to political pressures, import and export restrictions, and taxes. Lincoln used the argument of geographical unity and Manifest Destiny in an attempt to justify the war in his 1862 address to Congress. His argument employs the image of an abundant and economic nature:

> That portion of the earth's surface which is owned and inhabited by the people of the United States is well adapted to be the home of one national family, and it is not well adapted for two or more. Its vast extent and its variety of climate and productions are of advantage in this age for one people, whatever they might have been in former ages. Steam, telegraphs, and intelligence have brought these to be advantageous combinations for one united people.[5]

Following the reasoning of Manifest Destiny, it became every American's duty to convert wild nature into economic resources so as to improve the nation's industrial, communication, and transportation systems. Nature most benefited the individual and society when put to use to fuel the economy, civilize the continent, and build an empire. Prior to the mid-1800s, there was precious little talk about the need to conserve nature or prevent shortages of key resources. Conservation concerns, which now are commonly expressed in both public and private discussions, seemed irrelevant to a people faced with a vast, unexplored continent full of resources and endless riches. Now we are awash in pleas to reduce, reuse, and recycle. We have serious concerns about running out of oil, lumber, soil, water, fish, clean air, and other sources of energy, materials, and habitat that fuel the economy on which our well-being seem so dependent. Using nature is still important, but wasting it is less acceptable.

Nature as a Factory

A factory is supposed to be efficient and productive. Unmanaged nature may seem inefficient because it grows trees and other plants that don't have economic value, thus wasting the productive potential of soil and sun on so-called weed species that produce no profit. In addition, old trees are considered inefficient and something to cut and replace; they grow more slowly than younger trees, die, and rot before they can be converted into profitable lumber. An economic forest is an ecological machine managed for efficiency and profit. Only merchantable species are allowed to grow, and they are harvested before their rate of growth declines and well before they rot (see chapter 4).

Soil and forests were not the only things converted to profit by European settlers of America—whole species were nearly or completely extinguished in efforts to capture and maximize economic value. For example, millions of beavers and buffalos were trapped and shot; the economic value of their skins exceeded the value of their lives. These mammals that once dominated North American ecology nearly became extinct. Theodore Roosevelt (1901–1909) was perhaps the most conservation-minded president in history. He was an avid hunter and respected naturalist. By presidential decree, he created hundreds of millions of acres of national forests and preserves. This said, it is instructive to read the following quote that clearly reflects his economic construal of nature. By today's standards, the language is shockingly blunt.

Roosevelt knew that the vast herds of magnificent buffalo were disappearing from the American plains. Yet he sanctioned this loss because nature is first and foremost a factory fueling Manifest Destiny. It makes sense from this economic perspective to replace wild buffalo with efficient beef and to develop wild prairies into productive farms.

> While the slaughter of the buffalo has been in places needless and brutal, and while it is to be greatly regretted that the species is likely to become extinct, and while, moreover, from a purely selfish standpoint many, including myself, would rather see it continue to exist as the chief feature in the unchanged life of the Western wilderness; yet, on the other hand, it must be remembered that its continued existence in any numbers was absolutely incompatible with anything but a very sparse settlement of the country; and that its destruction was the condition precedent upon the advance of white civilization in the West, and was a positive boon to the more thrifty and industrious frontiersmen. Where the

buffalo were plenty, they ate up all the grass that could have supported cattle. . . . From the standpoint of humanity at large, the extermination of the buffalo has been a blessing.[6]

Livestock

Our language reflects and shapes our construal of nature. "Livestock" signifies that some animals are nothing but living capital, valued primarily for the profits they generate. We purchase hams, bacon, steak, chops, poultry, beef, and pork. Most of us don't see the connection between these refrigerated, plastic-wrapped pieces of meat and the warm, walking, living animals. Nonetheless, our economic decisions in grocery stores determine the fate of vast domains of nature.

The life of a steer provides a graphic example of nature defined entirely by the cold calculus of economics. Every aspect of a steer's life is determined by efforts to maximize profit. The steer's life begins when a breeding cow is artificially inseminated using sperm from a bull known to produce marketable beef. The steer will be yanked from the uterus, weaned as soon as possible, and pumped with hormones to increase rate of growth. He will be castrated and branded. After several months of open-field grazing, he will be penned in a small area for the reminder of his life, stand in his own feces, and unable to exercise. Grazing for grass is inefficient; it burns too many calories and slows weight gain. Instead, the steer will be force-fed a diet of corn and supplements that produce the marbled meat consumers prefer. Some of its food supplements are proteins from processed animal products that further speed weight gain. Thus beef eats beef (or horse, chicken, and sheep). This rich diet causes great distress to a multi-stomached digestive tract evolved to slowly digest grasses. Antibiotics and other medicines must be administered daily to prevent ulcers and infection.[7]

Even the duration of the steer's life depends upon economic calculations. In the short span of fourteen months, a newborn weighing eighty pounds rapidly gains over a thousand pounds. At some point, the accumulation of profit generated by the increment of weight drops below the accumulation of profit possible if money were invested in a bank or stock market, and the steer is more valuable dead than alive; the living stock is killed and converted into meat, which easily converts into money used to purchase baby cattle so that the process can be repeated and more profit generated.

Slaughter is designed to be quick, stress free, and relatively painless. The young steer will likely walk single file into a gently curving chute designed to obscure the view of other steers being killed. The lights will be dimmed to a calming level, and fans will blow from behind familiar smells from the stockyard rather than the frightening smells of slaughter just around the corner. A properly aimed pneumatic gun will cause instantaneous death by plunging a rod through forehead and brain. The act of slaughter may be one of the few aspects of the steer's life that is not determined entirely by economic rationale, although economic motives do exist to minimize animal stress. Animals scared into fighting or fleeing are difficult to handle, slowing down the assembly line; they also secrete stress hormones that degrade meat quality and sales price. However, the protests of animal-rights advocates have dramatically changed slaughterhouse practices. Slaughter often was protracted, painful, and gruesome. Some animals were bled alive so as to produce a more desirable and more easily handled quality of meat. Others were bludgeoned with axes without concern for quick death or minimal pain. Standards, imposed by mega-purchasers such as McDonald's in response to threats of consumer boycotts, now require that the animals die instantly and with minimal pain.

The economic efficiency of slaughterhouse practices is impressive. An economic use has been found for most every bovine part. Only 50 to 70 percent of the steer becomes food products. The skin, obviously, converts to leather. Other products are less obvious. Blood produces adhesives that bind the layers of wood in plywood used to build houses. It also gets used in fertilizers and foam fire extinguishers. Edible fats derived from bone, hooves, and horns are used in baked goods and chewing gum. Inedible fats are used to make plastics, tires, crayons, cosmetics, soaps, fabric softeners, and lubricants for jet engines. Cow connective tissues are used in numerous medical products, including dental implants and injectable collagen for plastic surgery. They also are used to produce jelly beans, margarine, shampoo, beauty masks, and photographic emulsion. All the organs are used, as well; for example, the lungs are used to produce a popular blood thinner and the pancreas is used for insulin. Little, if anything, is wasted.[8]

Land Conversion

The power of economics to define and transform nature can be seen as we glance at the landscape rolling by our car windows. A remote rural forest produces profit by growing fiber to make paper, timber to build houses, and deer to attract hunters. Trees may be replaced by sugar beets, corn, or cotton if these crops provide the owner greater profit. The forester turned farmer will be driven by economic calculus to plant every available acre with these valuable but expensive crops, apply fertilizer to spur their growth, and apply biocides to kill other plants that compete for the limited inputs of soil, sun, and water. The highest, best, and most profitable use of this once-forested land will change again as the tidal wave of human settlement rolls outward from urban centers. Roads, schools, and hospitals appear, and with them property values increase. Land planted in trees or crops produces greater economic value if replanted with houses. Croplands give way to curving roads, mailboxes, driveways, and suburban developments. Nature gets a manicure. Yards and potted plants replace forests and meadows. If the land happens to possess a mountain view or water frontage, then it might be sold again and that first, modest suburban house replaced with a larger and more expensive vacation home. Each transformation from forest to wealthy estate is driven by market values—nature gets transformed to produce the highest possible profit.

Genetic Engineering

New forms of life are created through genetic engineering. The process is time-consuming and hugely expensive. Not surprisingly, life-forms promising the most profit are the primary focus of commercial enterprises, perhaps making genetically engineered life the ultimate example of economic nature. Life-forms and processes that promise profit will be created, patented, and controlled by genetic engineering companies that base decisions on economic rate of return. Potential life-forms and processes that do not promise profit are far less likely to be of interest.

We need not venture into transgenetic technologies to illustrate how economics shapes life-forms. Take the common sheep as an example. Ten to twelve thousand years of selective breeding created sheep that are easily herded, possess thick wool, and quickly grow large, tasty flanks. Sheep have been changed by this process from an animal with a thin hairy coat, birthing one offspring per year, and lactating several gallons to something that annually provides ten pounds of wool, births multiple lambs multiple times per year, quickly adds weight, and produces gallons of dairy products. Today there exist nearly fifty sheep breeds of commercial value in the United States.[9] For example, merinos are valued for their fine wool, Lincolns for their coarse wool, and Suffolks for their meat and dairy production. These units of nature are now completely defined by wool production, dietary requirements, weight gain, milk production, and ease of butchering. Economic profit has directed sheep-breeding programs, so the essence of these animals has become aligned with economic criteria—their appearance, fat content, size, behavior, growth rate, and even reproduction.

SUSTAINABLE DEVELOPMENT

Sustainable development is an appealing concept because it merges two dominant but competing social paradigms: environmental protection and capitalist growth. The term's ambiguity creates problems if people who think they are talking about the same thing are actually seeking very different end results: sometimes it is not at all obvious what will be sustained and what will be developed. This ambiguity has advantages, however, because it can get people who would not otherwise talk with each other to sit around the table and look for common ground.[10]

Development, or growth, is embedded in the ethos of capitalism, the guiding philosophy of the global economy. The primary goal of a capitalist is to accumulate wealth, which means continual economic expansion so that there is more wealth to accumulate.[11] Few deny that economic growth affects environmental qualities. Capitalism converts landscapes to their highest economic potential, redesigns inefficient units of nature into factories, and eliminates the unmarketable parts of nature. But if capitalism has consumed nature, it also has delivered the goods. We have an abundance of choice and a quality of life unimaginable several generations ago. Because of these remarkable successes, and the promise of more to come, there is good reason to be cautious about policies that hinder economic growth. Advocates of capitalism acknowledge that growth may be uneven and messy but argue that it will eventually trickle down and improve the lot of the world's poor and hungry: the best way to cure poverty and inequity is not to divide a fixed pie into smaller pieces, but rather to grow a bigger pie through economic development.

Can development really be sustainable? Can economic growth fit within the constraints of ecological systems? Or will economic growth eventually and inevitably consume and degrade a finite planet? One argument promoting continued worldwide economic development is based on the famed Environmental Kuznets Curve.[12] This U-shaped curve proffers a favorable relationship between economic development and environmental quality. It postulates that economic development initially degrades environmental quality because pressures of establishing an industrial economy promote resource exploitation. But at some critical point the relationship inverts, and both affluence and environmental quality increase together. As economies mature, they undo the damage they wrought during their exploitative years.

The Environmental Kuznets Curve supposedly works because economic development changes people's priorities from survival to amenities. As they become more affluent, they turn their attention and money to creating environmental conditions in which they *want* to live. They also possess more technology and money to address the problems that concern them. As a result, technological and social practices emerge that enhance efficiency, reduce pollution, and produce amenities. According to some measures, advanced economies use fewer resources and generate less pollution to produce their desired goods and services. Empirical studies looking for evidence of the curve find it in some economies for some pollutants, especially air pollution. Evidence of the curve for many aspects of environmental quality, however, is lacking. One of the confounding factors is that developed economies often import their resources and manufactured goods, hence exporting resource exploitation and pollution. Improving environmental conditions also may require—in addition to economic wealth and technology—stable trading partners, enforceable regulations, minimal corruption, functioning democracy, and other factors that enable institutions to direct their attention to the environment.

So what is sustainable development? One frequently cited definition carefully avoids being too specific about the balance between sustaining ecological capital and sustaining social and economic capital: "Sustainable development is development that meets the needs of the present without compromising the ability of future generations to meet their own needs."[13] These future needs can be provided by either technological advances or functioning natural systems. Thus, following this definition, the conditions of sustainable development are met even if functioning ecological systems are degraded as long as social, economic, and technological capital exists to replace the values generated by the degraded natural systems.

The Environmental Kuznets Curve combines with this definition of sustainability to create the hope that the current system of continuous economic growth is not only sustainable but desirable. Together, they transform a zero-sum game into a win-win situation: economic growth supposedly makes the pie bigger and improves environmental quality. Thus policies such as trade liberalization, reduced government regulation, and global economy can be embraced because they promote capitalism and also happen to be good for the environment. Technological skeptics (chapter 4) question the ability of technology to create substitutes for all natural resources and ecosystem services. They worry that future generations will not be satisfied with the trade-offs we are making between increased economic development and ecological degradation because the proposed technologies may fail to deliver and because reliance on technological systems can restrict cultural opportunities by making future people reliant on sustaining the technology rather than the ecology.[14]

Hopes for continuous economic growth are critiqued by some environmentalists, who argue that capitalism will consume our habitat in order to fuel our economy. We will end up rich but dead. They argue that the endless pursuit of profit will eventually erode farmland, pollute water, deplete fisheries, drain wetlands, warm the climate, exhaust nonrenewable resources, and fill dumps. The philosophy of sustainable development implies infinite growth, and this seems incongruent with views from space of a beautiful and clearly finite blue-green orb surrounded by nothingness. They worry that every increasing disruption of ecological systems needed to fuel economic growth will eventually collapse the ecological interdependencies on which life depends.

The only solution, the critique continues, is to abandon the goal of economic growth. These environmentalists claim that "sustainable development" is a wolf in sheep's clothing, nothing more than a ruse promoted by greedy capitalists to co-opt and stall the environmental movement. Rather than strive for sustainable development, we should strive for a sustainable *balance* between humans and nature. Capitalism is flawed because it has no end point other than the complete domestication, consumption, and exhaustion of nature. This critique points out that the wealthiest 20 percent of the world's humans now consume 80 percent of Earth's resources and are responsible for equally disproportionate shares of pollution, global warming, deforestation, and desertification. What happens if poor economies prosper and all humans consume and pollute like typical Americans? Capitalism could consume dozens of Earths. Development cannot be sustained.

Some of today's business leaders agree with this critique and have responded constructively, recognizing that some current business practices are unsustainable. However, rather than abandon capitalism, they seek to modify it. These business leaders argue that we must work with economic markets because not only are these markets efficient; they also dominate our social and environmental institutions.

Ecological economics, green accounting, tradable emission allowances, green certification, cause marketing, and socially responsible investment funds illustrate just a few of the mechanisms that allow businesses to account for more than just profit. These modifications to capitalism allow consumers and investors to differentiate among costs and profits that degrade communities or enhance ecologies. Up until recently, the accounting system of capitalism was fairly crude. Success was measured as dollar profit versus dollar cost, and all transactions were reduced to this crude calculus. Some accounting mechanisms now allow businesses and consumers to distinguish how the profit was made and what environmental costs were incurred. It is becoming possible to tell whether a business practice exploits people, depletes resources, erodes soil, loses lives, creates inequity, or nurtures communities. In other words, it is becoming possible to discern the quality of profit, not just its quantity. Some call this a "restorative economy,"[15] aspects of which the following sections discuss.

PRICING NATURE'S SERVICES

New York, like most major cities, reaches well beyond its borders to obtain fresh drinking water. The city purchased rivers, forests, and farms; relocated cemeteries; and condemned whole towns so that vast acreages would collect, purify, and deliver water to the city. Ecological services performed at the Catskill and Delaware watersheds now provide about 90 percent of New York City's drinking water: evapotransportation essentially distills the water; precipitation delivers it to the region; soil microbes, aquatic plants, and tree roots break down and absorb contaminants; siltation traps heavy metals and floating debris; and gravity moves water through aquifers and along pipes. These free ecological services provide New York City the ample water of superb quality that has helped it grow in prosperity and worldwide stature.

However, land development, roads, septic tanks, car exhaust, agriculture, and industry were stressing the purification services. New York City considered installing a filtration plant similar to those of most other cities, at an estimated cost of $6 to $8 billion. With the bravado and creativity characteristic of New York, the city pursued an innovative alternative. The city believed it could protect the ecological services that purified its water for a quarter of the cost of artificial filtration. It did so by buying land to buffer against development and erosion, by upgrading sewage-treatment plants leaking into the watershed, and by encouraging land-development patterns and agricultural practices that minimize pollution and enhance water retention and purification. City residents transferred some of their wealth to rural communities, paying them for the ecological services their lands provided. The most profitable use of these lands changed from housing developments and industrial complexes to water storage and purification. As a side effect of these efforts, New Yorkers would also enjoy forested hillsides, green open spaces, wildlife habitat, and recreational opportunities.[16]

Nature is humanity's landlord, and we have been getting a great deal on rent and utilities. Oxygen is breathed but not paid for. Evaporation, rain, and gravity purify and distribute water. Microbes build and fertilize soil. Wastes are recycled or buried. Wild fish feed nations and natural forests warm them. Ozone blocks ultraviolet radiation. And insects pollinate food crops. The list of services that humanity accepts from nature goes on and on, and their combined economic value exceeds many trillions of dollars, surpassing by several times the gross national product of all countries combined.[17] Yet it is precisely these services that are degrading under the grinding weight of our consumptive lifestyles.

The more we learn about nature's services, the more we realize we lack the means to replicate them. Biosphere 2 provides a classic example. At the cost of several hundred million dollars, it was designed to help understand biosphere 1 (Earth) and to develop marketable technologies enabling extended space travel. Innovators sought to engineer a self-contained, human-controlled environment that reproduced the ecological services sustaining human life on Earth. Several people would live in it, growing all their food, producing all their own oxygen, and reprocessing all their wastes given only the input of sunlight. Filled with hope and promise, the well-funded project, staffed with the best minds, constructed and tested the impressive structure in the southern Arizona desert. Several acres were covered by glass domes interconnected by fans, pumps, pipes, wires, motors, and computers. Separate areas within Biosphere 2 replicated ocean, rain forest, forest plantation, and desert biomes. Although much has been learned about the technology needed to replicate nature, the engineered system failed on several occasions to sustain the lives of a few humans for a few months.[18]

Biosphere 2 teaches us that nature delivers many goods and services better and cheaper than human technology. It would be financially difficult and perhaps technologically impossible to engineer solutions to replace what exists in nature. Examples of underappreciated ecological services abound. Take a simple street tree, which, it turns out, does much more than provide roosts to songbirds and add color to our built environment. It shades our homes and streets, decreasing urban temperatures and moderating climate. It also absorbs carbon and pollution from the atmosphere, purifying air and water. It slows storm water runoff and thus moderates the need for concrete culverts, drains, ponds, and other expensive engineering solutions. Its beauty raises property values, attracts shoppers to stores, visitor to parks, and tourists to towns. When these and other benefits are priced, a single street tree can be worth a hundred times more than one of its rural cousins, well in excess of $1,000.[19] Many trees over a region can have far-reaching economic impacts. For example, trees strategically located to shade houses can reduce cooling needs to such an extent that fewer electricity generation plants are needed to power air conditioners during peak loads on hot summer days. Clearly it makes economic sense to maintain many of nature's services.

The dream of an emerging discipline named "ecological economics" is to produce "a world where Mother Nature at last receives fair compensation for her labor and recognition in our formal financial accounting" or, stated differently, to give Adam Smith's invisible hand a green thumb.[20] By identifying and quantifying nature's services, it is hoped that all environmental costs associated with supplying consumer goods and services will be reflected in the prices of these goods and services. For example, sulfur dioxide emitted from the smokestacks of coal-fired electricity generators causes acid rain, which sterilizes soil, kills fish, and weakens tree health. Thus, the electricity we consume with the unconscious flip of a switch has environmental costs that, up until recently, we have not been paying for on our electricity bills.

Innovative government pollution prevention programs set maximum sulfur-dioxide discharge quotas for each power generation facility. Newer, high-tech, cleaner-burning plants never reach their maximum loads and are allowed to sell their unused right to pollute to older, less-efficient plants. Thus, the price of clean electricity is lowered by profits from the sale of pollution permits, while the price of dirty electricity is raised by the cost of purchasing these permits. Ideally, these costs and profits get passed onto electricity consumers, who understandably prefer the less costly, cleaner electricity. In this way, the economics of pollution become internalized to business calculations. Inefficient power plants will gradually be retooled or replaced with efficient and less-polluting facilities because it makes economic sense to do so.

Ignorance of nature's services and the environmental costs of pollution present major obstacles to developing ecological economics, ignorance that only science can dispel. To give Adam Smith's invisible hand a green thumb, we need more knowledge about how human life depends upon nature so we can account for the ecological costs of our actions and include these costs into the prices we pay for goods and services. Currently we know little about these services, and even less about their market value.

GREEN CONSUMERISM

Take a walk through your local grocery store. Notice the advertising for natural cereals, biodegradable soaps, free-range eggs, recycled paper, and pure mountain drinking water. Compare their prices to products not engaged in green marketing. Why do people pay more for soup with no artificial flavors or for packaging made tree-free? Over at the pharmacy, you'll find natural remedies to cure what ails you. While driving home, you'll see more green advertising as you pass through neighborhoods called "Woodland Hills," "Evergreen Ridge," and "Deerfield Meadow." We spend billions of dollars each year visiting natural parks and gardens. We spend billions more surrounding ourselves with lawns, houseplants, picture windows, and images of nature. Obviously, many people value natural things, places, and products.

Some estimates of the "green market" suggest it includes more than one in every three consumers, captures several hundred billion dollars, and includes people willing to pay 5 percent to 40 percent more for green products proven to match or exceed the performance of nongreen products. The 1990s saw a dramatic rise in green marketing; one estimate suggests that 26 percent of all products launched in 1990 claimed to be ozone friendly, recyclable, biodegradable, or otherwise environmentally friendly. The market niche for "ethical" personal care products—those that don't exploit animals in product testing—is expected to grow 15 percent a year in the first decade of the twenty-first century.

The amount of green advertising tapered off somewhat after the mid-1990s as consumers became suspicious of ambiguous labels, false advertising, and poorly performing products.[21] Some marketing practices were justifiably labeled "green washing" because they misled consumers. However, several organizations are working to devise meaningful labels that make it easier for consumers to buy green. Several of these labeling efforts will be discussed later in the chapter.

What motivates a consumer to purchase green products? One reason is personal health. Consumers may hope that natural chemicals, pesticide-free products, and pure ingredients are less toxic and less harmful. Presumably evolution has tested these chemicals and humans have adapted to them. However, as reviewed in chapter 7, such assumptions can be misplaced. Nonetheless, many marketers have retargeted their green messages toward personal health, hoping the presumed link between nature and health will sell products to health-conscious baby boomers. Another reason people buy green is to save money. Appliances using less energy or water, for example, save the owner money over the long term even if the products cost a bit more to begin with. Likewise, building technologies such as solar heating and home insulation can reduce costs of heating, cooling, and lighting. Some products cost less because they use recycled parts (i.e., printer toner cartridges) or don't require disposal fees (i.e., when you pay someone to change the motor oil, tires, or batteries in your car, your bill often has an added fee to pay for disposing of the old parts).

Certainly one motivation for green consumerism is the desire to do something good for the environment. Most of us realize that our materialistic tendencies contribute to environmental degradation. We are well aware that the purchases making us happy, safe, beautiful, and plump consume scarce energy and materials and produce noxious pollution. It is understandably tempting to alleviate guilt by purchasing things good for nature. "Sustainable consumption," or "shopping for a better world," seemingly allows us to have our cake and eat it too. Both the United Nations and the U.S. Environmental Protection Agency have programs to assist consumers in making green purchases. The UN program even goes so far as to suggest that "Sustainable Consumption is not about consuming less, it is about consuming differently, consuming efficiently, and having an improved quality of life."[22]

Critics claim green consumerism is dangerous and misleading. It just reduces feelings of guilt and encourages our materialistic orgy rather than advancing the necessary but harder-to-sell policies of reducing, reusing, and recycling. Proponents counter that green consumerism is one way individuals can make a difference (most of us must make purchases; we might as well make green ones) and collectively our decisions can motivate business practices that promote sustainability. Certainly, if major institutions such as the U.S. government and the United Nations implement environmentally sensitive purchasing programs, as they have promised to do, then businesses will respond, creating an array of green products from which the rest of us can also choose. Some visions of a green economy see product design and manufacturing systems retooled to use mostly materials recycled from returned products. The waste of any product or production system becomes the nutrient for another product or system. Consumers rent rather than purchase products and return them for newer versions and more services.[23]

One of the main challenges to green consumerism is identifying green products so that consumers can make green purchases. While strolling through most shopping centers, you may purchase a variety of products labeled as "natural": natural formula laxatives, condoms made of natural latex, cigarettes with natural menthol, and natural-smelling air fresheners. What does it mean to label something "natural"? Maybe not that much. Many labels exist, but meaningful, sincere, and understandable ones are few and far between.

Misleading product labeling is one reason green consumerism lost some momentum in the late 1990s. In one rather infamous example, a major discount retailer labeled a brand of paper towels "green" simply because the interior roll on which the paper towels were wrapped was made from partially recycled fibers. Neither the paper towels nor the plastic wrap in which they were packaged contained recycled materials. Was this label misleading?

After numerous revisions, the food-labeling system is becoming familiar, accepted, and effective at communicating the calories, fats, vitamins, and other diet-relevant information derived from each serving of a product. But environmental labels have a long way to go. It is particularly difficult to communicate whether a product is good for the environment because there are so many different types of environmental impacts. For instance, labeling will need to communicate information about whether product development abused animals, the raw materials were harvested sustainably, manufacturing was nonpolluting, packaging material is recycled, distribution to retail outlets is energy efficient, use of the product saves energy or causes health risks, the product is reusable, and the company's labor practices are socially responsible. Obviously there exist many opportunities for labels to mislead and confuse consumers.

Several labeling systems are competing to develop a comprehensive environmental rating for every product. Some of these systems employ "life-cycle" analysis, the goal of which is to assess the environmental impacts that accumulate from cradle to grave—from the birth of the product, through its use, on to its disposal, and ideally back again into reuse. Not surprisingly,

Table 1: Green Product Labels

Type of Message	Example of Labels
Human health	Organic, pure, pesticide free, nontoxic
Animal rights	Free range, no animal testing, cruelty free
Save money	Energy star, energy efficient, water miser
Save the environment	Recycled, reusable, tree-free paper, biodegradable, ozone friendly, sustainably harvested, nonpolluting
Ambiguous	Natural, 100% pure, Earth friendly, bio-based, eco-safe, Earth smart, green

such analyses are incredibly difficult and often contested. Take a ceramic coffee mug as an example. Should we begin our analysis at the site where materials are harvested, or should we begin with analysis of the machinery used to harvest the materials? Should we end the analysis assuming all consumers are equal in their care for the cups, or should we differentiate between those who save water by washing only with full dishwashers set on "water miser" and have very sure hands that never drop and break cups?

The choice between disposable and cloth diapers was an early test case for life-cycle analysis. Disposable diapers are made of nonrenewable resources such as plastic, as well as renewable paper fiber. Cloth diapers are made of renewable materials and are reusable after washing. Growing cotton and washing diapers both require abundant amounts of water, perhaps using four to six times more water than used in manufacturing disposables. Cloth diapers also require more energy. Most of the cotton is shipped from the United States to Asia for manufacturing and back to the States for sale. It also takes more energy to plant and harvest cotton than trees, not to mention the energy required to launder diapers. Both tree plantations and cotton fields are fertilized and sprayed with pesticides, but growing cotton requires a more intensive use of these toxins. Laundry detergents and bleach add to the toxin load of cloth diapers. Disposable diapers place fecal material in landfills, while cloth diapers place it in sewage systems. Disposables, obviously, require more landfill space. On and on we can go.

Unfortunately, no clear choice emerges between disposables and cloth diapers, even at this level of analysis. But the analysis is still useful. In regions short on water and flush with landfill space, disposables might be more appropriate. In regions flush with water but short on landfills, cloth diapers might be more appropriate. Not all life-cycle analyses are unequivocal. The choices between recycled or virgin aluminum, for example, are more obvious because of the enormous energy required to extract aluminum from virgin ore.

As a consumer, you know that labels can be vague if not deceptive. The Consumers Union, the organization that reviews retail products and publishes *Consumer Reports,* has a website that evaluates environmental labels, most of which it finds to be inconsistently applied, difficult to verify, and not very meaningful.[24] Few government regulations exist, so most labeling is voluntary. Progress is being made, however, by motivated people devising ways to green the economy. Product labels such as "Green Star," "NutriClean," "Energy Star," "Forestry Stewardship Council," and "ISO 14001" are increasingly gaining industry respect and consumer confidence because they are administered by a third party, independent of the business interest.

Labeling is only one cog in the wheel of a green economy. A product's environmental impacts depend upon how the product is used as well as how it is manufactured, transported, and disposed. Ceramic coffee mugs, for example, may appear more sustainable than disposable plastic cups because silica can be safely harvested from abundant sources and mugs can be used countless times. However, the enormous energy required to fire the ceramic and transport the heavy mug to market may outweigh any environmental benefits unless the mug is reused at least a thousand times. If the mug cracks beyond repair during its nine hundredth washing, for example, then it might have been less environmentally destructive to use Styrofoam cups.[25] Even energy-efficient fluorescent bulbs might be less environmentally sound than old-fashioned incandescent bulbs if installed where they will be frequently switched on and off. In such situations, fluorescents use more energy to operate and wear out faster, not justifying the extra materials to manufacture and the weight to transport.

GREENING BUSINESS

Responding to increased consumer demand is only one reason businesses' might act "enviropreneurially."[26] Other pragmatic reasons exist. For example, businesses must respond to government regulations. Early twentieth-century regulations required businesses to protect worker health and safety. Cities increasingly regulate land uses to reduce threats to human health and well-being. And federal and state regulations further restrict business practices that dump hazardous wastes and other pollutants into the public's air, water, soil, and food. Fines for violating these regulations eat into corporate profits.[27]

The cost of environmental cleanup and liability provides another motivation for greening business practices. Several examples should suffice to make this point. The 1984 accidental release of chemicals by a Union Carbide plant in Bhopal, India, killed several thousand people and injured perhaps several hundred thousand more, resulting in approximately 145 legal actions involving 200,000 plaintiffs and millions of dollars in settlements. Illness caused by coal dust and asbestos produced dozens of lawsuits and cost companies billions of dollars. And the oil spill in Valdez, Alaska, cost Exxon several billion dollars in cleanup costs, damage settlements, and penalties.

Insurance and lending institutions are taking notice because these expenses affect company profit and solvency. The U.S. insurance industry, for example, faces $2 trillion in claims for asbestos and other pollutant cleanups and, according to one estimate, spends about $450 million a year just on transaction and legal costs associated with sites designated by the U.S. EPA as toxic and in need of remediation. By necessity, insurers have begun to motivate environmentally sound business practices by refusing to insure or charging higher premiums on environmentally risky businesses. The lessons are as obvious as they are expensive: products, practices, and pollutants that threaten human health and environmental quality create expensive liabilities. It pays to be green.[28]

Environmentally destructive business practices run the risk of becoming targets of advertising campaigns and boycotts that damage business reputations and reduce profits. These campaigns are often launched by environmentally motivated nongovernmental organizations and can have dramatic effects on sales. Dolphin-safe tuna is perhaps the most famous example. Tuna swim below dolphins, so commercial tuna boats would cast their nets where dolphin swam and the dolphins would drown in these nets. Greenpeace filmed dead dolphins being thrown overboard from tuna trawlers and mounted a public relations campaign that motivated major tuna fish retailers to insist that their tuna suppliers change fishing practices. Once one company promised to buy and sell only dolphin-safe tuna, other retailers followed suit and the whole tuna-fishing industry changed. The number of annual dolphin mortalities dropped from 300,000 to 3,000 in just a few years.

In response to similar media campaigns and boycott pressures, McDonald's, Burger King, and other large buyers of beef and poultry have imposed strict animal handling guidelines and audits to ensure humane animal treatment. The first guidelines insisted on quick and painless death at slaughterhouses, then increasingly included many aspects of an animal's life, including castration, debeaking, dehorning, branding, ear notching, thermal comfort and air quality of animals in closed environments, forced starvation, and methods of transportation. In response to mounting concerns over human health risks of excessive exposure to antibiotics in meat products, McDonald's has recently asked its meat suppliers to phase out use of antibiotics routinely given to animals to promote growth. The policies of these large corporations, with huge supply chains, have forced dramatic changes in on-the-ground management of nature.

The largest home-improvement chain store in the United States, Home Depot, provides another dramatic example of economic motivations for greening businesses. It sells thousands of building materials, some of which used to be made from endangered, ancient, or unsustainably grown trees. Pressure from environmental groups combined with a desire to protect corporate image and profits motivated this retail giant to establish green-purchasing polices. Home Depot specifically pledges to "stop selling wood products from environmentally sensitive areas" and to be "a global leader to help protect endangered forests." This mega-retailer further pledged to stop selling wood products made from endangered or ancient trees and to increasingly stock wood products that contain recycled material and/or are certified by a third party to be grown sustainably.[29]

Changes made by these large companies have profound impacts because they ripple down the supply chain and throughout the economy, forcing competitors to change in order to stay competitive and suppliers to change if they want to continue supplying the large companies. The ripple effects of large corporations are extensive, and their actions can affect whole market sectors. For this reason, sophisticated nongovernmental, governmental, and business activists increasingly advocate green supply chains. Levi Strauss, Xerox, Procter & Gamble, and 3M are just a few of the multinational corporations requiring their suppliers to meet strict environmental criteria in addition to those of price, quality, and delivery standards.

Redesigning facilities and production processes with an eye toward eco-efficiency can also make money in immediate and obvious ways. Environmental Audits and Design for the Environment programs show companies how saving energy, recycling wastes, reducing pollution, and other environmentally friendly practices actually save money.[30] For example, Lockheed built a 600,000-square-foot structure using environmentally friendly designs that utilize sunshine for most lighting needs. The building modifications cost an extra $2 million. The payback in energy savings, however, is close to $500,000 per year. Worker productivity was also enhanced.

Planning for product disassembly or recycling saves money by reducing the need to purchase new supplies. If products at the end of their useful life are returned, disassembled, and partially reused, then businesses can save on production costs of new products. Republic Engineering Steel, for example, saved nearly $3.6 million a year by recycling its steel more efficiently. In a related policy, many hotels now save thousands of dollars monthly on energy and water costs by allowing patrons to decline daily laundering of towels and linens.[31]

Access to capital provides yet another reason for green business practices. Socially and environmentally responsible investing directs a small but influential proportion of stock market investing. Specific funds screen companies using environmental practices as criteria. Perhaps more influential are large investment agents such as teacher retirement funds and state pensions. Managers of these institutions are coming under increasing pressure from their shareholders to invest in socially and environmentally responsible businesses. These fund and pension managers, in turn, apply considerable pressure on companies to green-up their practices. Publicly traded corporations also come under pressure from environmentally motivated shareholders who sponsor resolutions forcing green business practices, such as subscribing to the principles of the Coalition for Environmentally Responsible Economies (CERES). CERES established ten guidelines for environmentally responsible business in 1989. They enlisted support from investors, advisers, and analysts with influence over $300 billion of investment capital, as well as environmental and public interest groups, to encourage companies to adopt these principles. Over seventy companies have signed on, including large multinationals such as American Airlines, Bank of America, General Motors, and Polaroid. Investors, chief executive officers, and financial analysts now realize that environmentally sound companies are also well-run companies positioned for leadership. A number of studies in the late 1990s demonstrated that companies outperforming their peers environmentally also outperformed them in the stock market: they were less-risky investments, enjoyed a lower cost of capital, and earned a higher stock price. Being green turns out to be good for business.[32]

Green business practices are motivated by business ethics as well as by profit motives. Many business leaders believe acting environmentally responsible is the right thing to do. Businesses are the dominant institutions in the global economy, and hence among the most powerful institutions on Earth. Some business leaders recognize their social role and accept that their policies affect the fate of Earth and humanity. Astute leaders see that desertification, global warming, biodiversity loss, resource depletion, and general environmental degradation cannot be good for long-term prospects of business, especially to the extent that these problems promote famine, disease, and social unrest. Moreover, all businesses must function within the bounds of acceptable practices defined by citizens and consumers. Successful consumer boycotts and pressures from socially responsible investing demonstrate how this social contract is changing. Thus, both selfish and ethical reasons motivate leaders of such notable multinationals as Chevron, Volkswagen, Nissan, Mitsubishi, Dow Chemical, Alcoa, and Shell to sign the Declaration of the Business Council for Sustainable Development, which begins with the following words:

> Business will play a vital role in the future health of this planet. As business leaders, we are committed to sustainable development, to meeting the needs of the

present without compromising the welfare of future generations. This concept recognizes that economic growth and environmental protection are inextricably linked, and that the quality of present and future life rests on meeting basic human needs without destroying the environment on which all life depends.[33]

In addition to the Business Council quoted above, numerous other institutions exist with similar purposes, such as the Global Environmental Management Initiative, the Alliance for Environmental Innovation, and the International Chamber of Commerce's Business Charter for Sustainable Development. Successful business leaders have begun advocating policies within their companies that promote sustainable development. Patagonia and the Body Shop are often held up as poster children of green businesses, but credible efforts by multinationals such as McDonald's, Xerox, and 3M demonstrate that they are also leaders in reducing waste, pollution, and animal cruelty.[34]

Certainly the road ahead is not clear of obstacles. The color of all businesses is not green. Greed and the need to maintain cash flow still drive many business practices. And even so-called green business practices and declarations of sustainability need to prove that they are, in fact, making a difference. However, the old harsh axiom that the business of business is business may be softening with awareness that the goals of good business and social good are interwoven.

CONCLUSION

Many people enjoy comforts, safety, and conveniences far in excess of what royalty experienced just a few generations back. Capitalism has indeed delivered the goods. It also has created potential problems. The improved living conditions are so unevenly distributed that many humans and many other creatures live in impoverished conditions. Correcting these inequities is only part of the challenge. The long-term challenge of balancing ecological sustainability and economic development is daunting. Enormous obstacles and much skepticism remain about whether capitalism can do anything but consume a finite Earth. The dominance of business institutions, however, means that ecological economics and green business practices are likely to be part of any credible effort to achieve sustainability, even though skeptics point out that economic development and the institutions that support it (science, technology, expertise, large hierarchical systems, and the values of efficiency and growth) are what created many of the so-called environmental problems in the first place.

Green capitalism is understandably attractive to political, business, and dominant social institutions as a solution to environmental problems.[35]

It emphasizes the free market over government regulations, which is consistent with deregulation and decentralization tendencies started in the early 1980s and continuing into the twenty-first century. By promising to grow the pie bigger, it transforms the environmental problem from a zero-sum game where some constituents must lose into a win-win situation where many constituents can win. It promises increased profits, new markets, added value, and satisfied voters. Finally, as noted, green capitalism employs the techno-scientific institutions that are already well established in the corridors of power.

Despite this faith in capitalism and economics as means to cure social and environmental ills, we cannot escape the fact that we live on a finite world, with a continuous but finite flow of sun, gravity, matter, and other energy sources. These limits did not restrict choice when the world was sparsely populated by humans, but now Earth is relatively full of us and our doings. Our materialistic culture places increasing demands to convert finite nature into economic resources by encouraging us to satisfy urges and build identity through consumerism. Our desire for excess might doom us unless and until we can define the good life in such a way that rewards personal fulfillment at least as much as materialistic accumulation.

Even if economic systems can be tweaked to ensure sustainability, we may not like living in an entirely economic nature and thus might want to be cautious in how far we follow the economic calculus. Ecosystem services on which human life depends (such as water purification) may fare well using economic criteria to guide land development, but religious, aesthetic, bio-rights, educational, and other values may not. Inefficient and unmarketable qualities of nature will be reengineered, neglected, or replaced. The Chesapeake Bay, for example, provides economically valuable services as a liquid highway for shipping, a port for the maintenance and construction of navy and other ships, a sewer for communities along the coast, and a slippery surface for recreational motor boating at high speeds and loud volume. Of less economic value are the oyster- and fishing-dependent rural communities destroyed by siltation, historic settlements of Native Americans and European pioneers converted to suburbia, and even the internationally famous blue crabs losing their habitat. Species, lifestyle, and history do have some economic value, but nothing compared to the billions of dollars derived from industry, sewage treatment, and motor-based recreation. Thus, many of the qualities that currently define the bay and motivate our concern for it might not fare well if we base management only on the most efficient means to turn a profit.[36]

* 7 *
Healthy Nature

Physical Health · Emotional Health · Risk

Rachel Carson's 1962 landmark book *Silent Spring* begins by describing a small town in the heartland of America—once known for its charming landscape, happy children, and thriving farms—that had fallen victim to blight and stillness. Children became ill, chicken eggs no longer hatched, and the vegetation browned. Spring was silent because birds had died or left. The cause? Pesticides sprayed indiscriminately in an unnecessary war against insects. The book struck a chord in the American consciousness: it became a bestseller, the chemical industry mounted a dramatic and expensive counteroffensive, *60 Minutes* aired a segment on the problem, and Congress held hearings. Carson argued forcefully for constitutional protection against the hazards of chemical pollution:

> We have subjected enormous numbers of people to contact with these poisons, without their consent and often without their knowledge. If the Bill of Rights contains no guarantee that a citizen shall be secure against lethal poisons distributed either by private individuals or by public officials, it is surely only because our forefathers, despite their considerable wisdom and foresight, could conceive of no such problem.[1]

Silent Spring changed the focus of environmentalism. It showed that more was at stake than saving finite resources, charismatic wildlife, and aesthetic wilderness experiences. Economic growth and technological arrogance put our minds and bodies at risk. This period in history has been heralded as the birth of the modern environmental movement. Public opinion polls during the next few decades showed environmental issues rivaling jobs, taxes, and national defense for public attention. Membership in environmental organizations mushroomed, from tens of thousands to tens of millions. So did environmental regulations, increasing from dozens to nearly a hundred thousand by the end of the century.

This chapter explores three links between nature and human health that dominated public understanding of environmental issues: (1) clean air, clean water, safe food, and related factors that affect human health; (2) nature as a remedy for the stress and anomie of living in an increasingly built and crowded world; and (3) using risk of ill health or death as a way to understand environmental quality.

ENVIRONMENTAL LINKS TO HUMAN PHYSICAL HEALTH

Rachel Carson was not the first to rally environmentalists around human health. Equally acute concerns erupted in the mid-1800s as urbanization and industrialization mixed people's air, water, and food with smoke, sewage, and hazardous wastes. Urban air was often acidic and thick with particulate matter. Residential and industrial wastes polluted rivers, lakes, wells, and other sources of drinking water. Yellow fever, cholera, and typhoid plagued industrializing cities. In 1854 John Snow, a London physician, was the first to demonstrate the link between cholera outbreaks and human sewage leaking into water supplies. Other studies followed to show additional links between environmental quality and health.

Civic organizations pushed for improved environmental conditions, but progress was slow. The first municipal sewage-treatment systems, for example, merely drained raw sewage in open ditches away from residences and water supplies, dumping it into the rivers or oceans. Eventually open ditches were replaced by pipes and, much later, the pipes led to treatment systems. The earliest such systems merely settled the solid waste before releasing the rest. It was not until the middle of the twentieth century that progressive cities such as Boston, New York, Chicago, and Los Angles used sand filters and additional treatment steps before releasing waste water. Meanwhile, cities built another pipe system to deliver fresh water from reservoirs located upstream of sewage discharges. Control of industrial toxins happened much later. Environmental cleanup was slow then, as now, because of concerns that pollution controls would harm economic prosperity. However, increasingly activist governments gradually established institutions such as boards of health, which led eventually to the establishment of the U.S. Public Health Service in 1912.

Social activists and environmental professionals coalesced around the need for healthy food, fresh air, potable water, and safe working conditions. Many of the environmental sciences and professions we know today emerged in response to pressing concerns about pollution. Cities gradually controlled the discharge of wastes and found additional ways to secure the air, water, and food needed to sustain and attract industry, workers, and taxes. Unions and government regulations worked to eliminate practices that exposed workers to toxic chemicals and lethal conditions. The political momentum from this urban-centered, turn-of-the-century movement probably transferred to rural environmental concerns and enabled creation of the U.S. Forest Service, the National Park Service, and other institutions charged with addressing forest, park, wildlife, wilderness, and related natural conservation problems.[2]

Several generations later, Carson's *Silent Spring* refocused public attention on health, only her story emphasized something more sinister than sewage: the ubiquitous and invisible pesticides being aggressively promoted as tools to control nature and improve quality of life. In particular, she focused on DDT, a cheap, effective, and long-lasting chemical that kills many species of insects such as lice, ants, and mosquitoes. The first major use was by the U.S. military to control lice that spread typhus and other diseases among troops. After World War II, U.S. production of DDT reached fifty thousand metric tons annually, almost one pound per person. Cheap surplus war aircraft sprayed the effective but inexpensive chemical over many North American farms, forests, and housing developments in attempts to control "pests" that destroyed crops, harmed livestock, transmitted diseases, and annoyed tourists. Applications of DDT in malaria-prone regions of the world saved millions of people from mosquito-transmitted malaria.

The widespread spraying also killed the "good" insects, as well as some of the fish, birds, and other animals further up the food chain. Mammals resist DDT's toxicity but readily store it in their fatty tissue. It does not easily degrade and thus remains effective for over fifty years. In fact, it is still found in and around our homes in quantities greater than many modern pesticides; virtually all humans born after 1940 have DDT in their fatty tissues. A chemical revolution occurred about the time of World War II, rivaling the Industrial Revolution in scale and importance. Pesticides, drugs, solvents, paints, and preservatives changed the way Americans lived. The amount of synthetic chemicals produced in the United States rose from 1 billion pounds in 1940 to 30 billion pounds in 1950 and 300 billion pounds in 1976.[3]

Carson's alarming scenario of a silent spring struck a chord in a public already anxious about their well-being. The cold war, fallout shelters, and civil defense warnings had created a palpable fear of nuclear winters, radiation poisoning, and a dread that humanity might destroy itself. These silent, invisible killing chemicals were being found in the world's most remote locations as well as in the very food we eat. In the 1950s, news reports warned that radioactive iodine from atomic blasts had been detected in locations as remote as Alaska. Also in the 1950s, fruits and vegetables such as cranberries sold for human consumption were publicly recalled because of pesticide contamination.

Events following publication of *Silent Spring* fueled the flame of public concern. In 1964 the U.S. Public Health Service attributed a massive fish kill in the lower Mississippi River to a pesticide discharge in Memphis, Tennessee. Two years later approximately eighty deaths were attributed to an air inversion that concentrated pollution in New York City, and public health officials estimated that less acute air pollution events in cities around the world caused many more deaths and illnesses each year. In 1969 the Cuyahoga River near Cleveland burst into flames due to petrochemical runoff. By the early 1970s, Lake Erie was declared dead and a place "where fish go to die" (it eventually recovered, but new dead zones appeared in 2002). Meanwhile, lead, invisible but omnipresent because it was added to paint and gasoline, was weakening our children, degrading their central nervous systems, killing off kidney functions, and dulling intelligence. Even the ability of the atmosphere to protect us from radiation faltered. In the 1970s we began to hear about the thinning ozone layer. The chlorofluorocarbons we innocently used for refrigeration and spray-can propellants decompose into chemicals that destroy the ultraviolet radiation-absorbing ozone molecules in the stratosphere. The increased radiation causes eye damage and skin cancer.

Carelessness added insult to injury, providing reason to doubt humanity's ability to deal with the gifts inside Pandora's box. Reports about mistakes and ineptitude appeared in the news with alarming frequency. For example, in 1973 an inspector at the Hanford Nuclear Reservation near Richland, Washington, went on vacation for several weeks. An improperly trained replacement allowed 115,000 gallons of highly radioactive liquid wastes to escape and potentially leak into the groundwater used by the local community. At about the same time, a thousand miles east, it was discovered that a wood preservation facility near Laramie, Wyoming, had "lost" huge quantities of wood-preserving toxins such as arsenic. The hundred-year-old facility soaked wood in open pits filled with the toxic chemicals that killed insects, rot-causing bacteria, and just about everything else. Amazingly, no one thought to be alarmed that 100,000 gallons of this material leaked into the soil each year (remedial action began in 1983). As horror stories of this kind continued, the link between the environment and human health came to the fore of public awareness.

Love Canal represents the epitome of public concern. In the 1970s residents near Niagara Falls, New York, complained about rashes and eye irritation. Not much was done about these complaints until 1976 when the International Joint Commission on the Great Lakes sought to find the source of a banned pesticide that was poisoning fish in the Great Lakes. They traced the flow of the toxin up through the water system and ultimately to the Love Canal neighborhood.

A bit of historical research revealed that between 1942 and 1953 a chemical corporation sealed its chemical wastes in steel drums and dumped them in a nearby pit. In 1953 the pit was covered with clay and topsoil and sold to the Niagara Falls school board for $1. Some years later an elementary school, playing fields, and 949 homes were built where the chemical company once stood. The drums eventually corroded and their contents began leaking into storm sewers, gardens, basements, and even the school playground.

This revelation attracted considerable media attention. Public meetings, scientific studies, distrust, and recrimination followed. Residents expressed concerns not only about bad smells and rashes but illnesses such as birth defects, miscarriages, assorted cancers, and respiratory disorders. In 1978 New York State closed the school, permanently relocated 238 families closest to the dump, and fenced off the area. President Carter declared the whole Love Canal site a federal disaster area and had the remaining families relocated. Eventually the dump site was covered with a new clay cap and surrounded by a drain system that pumps leaking wastes to a treatment plant. By the 1990s the cost for cleanup and relocation had reached $275 million. The area, renamed Black Creek Village, was declared "habitable." People now live nearby.

Subsequent studies have not found increased incidence of cancer in Love Canal residents compared to those living in the larger Niagara Falls region, but those families living closest to the dump did have a higher incidence of miscarriages and low birth weight. The link between toxins and human health remains controversial, but the Love Canal and similar incidents located around the nation created a public awareness that maybe no one and no place was safe—the chemical by-products of our industry may literally be poisoning our own backyards.[4]

Current debates over pesticide use and food safety reflect public concerns about the link between human health and environmental quality. Pesticides are designed to be harmful to some organisms but not to others. They are so widely applied as to be everywhere in the biosphere. People get exposed through inhalation, skin absorption, and ingestion. Most of the foods we eat are grown using pesticides, so their residues may be uninvited guests at our dinner tables. Many of us also use these chemicals in and around our home to control insects, weeds, mold, mildew, bacteria, and to protect our pets from ticks and fleas.[5]

The public's primary concern about pesticides seems to be the health risks associated with residue in food. Some chemicals get stored in animal fat rather than being excreted and thus accumulate in us over time. Their concentrations increase as we look further and further up the food chain because predators accumulate the toxins of their prey. Eating near the top of the food chain, humans may be particularly vulnerable. Some of these chemicals even appear in human breast milk near toxic levels. These chemicals evoke great passion and concern because we know they exist, we know they accumulate in our bodies, but we have no way to observe them without sophisticated equipment, and even then it is difficult. We worry about toxic time bombs ticking away in our bodies. Exposures earlier in our lives may accumulate to a critical mass and produce cancer or some other illness. The stealth and mystery of these chemicals makes them seem sinister. But is our toxic load too much? How much is too much?

Some of us respond to these concerns by purchasing "all-natural" or "pesticide-free" foods. The organic food industry has grown to a multibillion-dollar business, expanding 20 percent a year between 1990 and 2000. Forty-three percent of American consumers say they would like to see "organic" food production become the dominant form of agriculture in the United States.[6] People seem to assume that "natural" is safer than synthetic, even though there is ample evidence that natural toxins can be just as problematic. For example, a fungicide EDB was used to kill fungus that grows in nuts and grains. EDB is carcinogenic, but so is the aflatoxin produced by a fungus that thrives in grain storage silos. In fact, aflatoxin is among the most potent carcinogens known, much more so than EDB. When EDB was banned as a pesticide, the rates of aflatoxin increased, quite possibly exposing the public to a much higher risk of cancer. Pesticide-free corn and peanuts can be similarly contaminated without careful grain storage techniques. Food preservatives, milk pasteurization, meat irradiation, and related techniques were developed to control and remove hazards from our food system. Thus, the organic and natural labels do not always translate into healthy and safe foods.[7]

Pesticide technology continues to advance. The very early pesticides such as arsenic and mercury affected a broad class of organisms; they effectively destroyed pests but also were toxic to the applicator and most everything else in the environment. Inventions such as DDT were far less toxic to nonpests and cheaper to produce. But because they were long lasting (originally thought to be a good thing because fewer applications were required), they spread great distances after application and accumulated in all life on Earth. Technological advances allow newer pesticides to target only specific pests. Unlike DDT, they are intentionally designed *not* to be lethal to broad classes of life and to decay quickly. Exposure to light, water, and time degrade their chemical compounds into nontoxic components. As a result, pesticide residue, if present in our food and water, is much less likely to be toxic.

Humans get exposed to thousands of organic and synthetic chemicals other than those consumed through the food we eat and the water we drink; many pose potential health threats. For example, there is currently some concern about lung or bladder cancer caused by arsenic absorbed when children use playground equipment built of chemically treated wood.[8] Apparently, arsenic on the wood's surface gets picked up on the hands of children and transferred to their mouths during and after playing. Risks vary with the type of wood and amount of playing, and can be controlled by thorough hand washings with soap and water. However, everyday situations give people reason to worry about their toxic loads.

These pesticides and other toxins are only part of the reason people fear their health might be in danger. Rarely a day goes by without us hearing about global warming, acid rain, soil erosion, water shortages, destruction of the world's forests, or some other kind of environmental degradation. In addition to worrying about running out of resources, we now also worry about complete and widespread ecological collapse. Species extinction and the consequent loss of biodiversity provide just one example, as E. O. Wilson explains:

> ... the worst thing that will probably happen—in fact is already well underway—is not energy depletion, economic collapse, conventional war, or even the expansion of totalitarian governments. As terrible as these catastrophes would be for us, they can be repaired within a few generations. The one process now ongoing that will take millions of years to correct is the loss of genetic and species diversity by the destruction of natural habitats. This is the folly our descendants are least likely to forgive us.[9]

Thousands of species have been extinguished by human action. Some estimates suggest that thousands more go extinct each year, creating the greatest mass extinction in evolutionary history. Yet it is not at all obvious the degree to which these extinctions will impact human survival. The basic services on which life depends seem to continue despite the loss of numerous species. When one species goes extinct, others seem to move in and assume its ecological functions. Nature, fortunately, seems to have multiple redundancies. But how far can the system be pushed before these redundancies are exhausted? Many people are concerned but no one knows the answer. Consider the following analogy to an airplane: If the species going extinct are just passengers in an airplane, then the plane keeps flying. If the species are the millions of rivets holding together the fuselage, then we can lose quite a few species without worry. However, at some point we are likely to remove enough strength from the structure and lose a tail or a wing or some other essential part. If a species happens to play a critical role, such as the airplane pilot or engine oil seal, then even the extinction of that one species may cause a crash.

There are indeed many links between environmental quality and human health that shape public understanding of environmental issues. The obvious threats such as sewage and factory wastes spewing untreated from pipes into drinking water supplies were easy to identify and fix. Invisible pesticides and other toxic by-products of the chemical revolution require more sophisticated responses, but at least the causes are known. The health implications of biodiversity decline and global climate change are particularly difficult to estimate because the risks are unknown and the causes are inextricably linked to our lifestyles. Potential new threats such as those posed by Frankenstein-esque genetically modified organisms seem even more sinister because engineered life can reproduce and change itself, escaping any hope of human control.

PSYCHOLOGICAL HEALTH

Before shifting attention to the question of risk and determining how much risk is too much, we should consider that the health effects of environmental quality extend beyond physical health to include emotional and mental well-being. One of the best-known studies of this link between nature and health found that surgery patients recovering with views of nature require less painkilling medication and were discharged sooner than were patients with similar doctors and similar surgery but without a view of nature. Nature seems to relax people, reducing and even inoculating them against future stress. Experimental subjects exposed to stressful situations such as viewing hip-replacement surgery experienced less stress if they were shown pictures of natural areas prior to or just after the stressful film. People shown pictures of modern art or automobile traffic experienced more stress in these controlled and replicated experiments. Numerous other studies demonstrated that encounters with nature significantly reduced blood pressure, slowed heart rate, and lessened stress-related steroids present in blood and saliva. These encounters with nature also reduced subjective reports of stress and improved moods and feelings of vigor. In a health-conscious nation, such evidence provides powerful justification for protecting nature.[10]

Advocacy for the health-promoting effects of nature existed long before scientific study proved the relationship. The rapid change from agrarian to industrial lifestyles created worries about the long-term health consequences of crowding, traffic congestion, constant noise from motors and speakers, the stench of smoke, and the general strain of living in a concrete jungle without access to nature or solitude. Engineering solutions (i.e., mufflers for motors, pollution controls of tailpipe emissions, more roads for faster commutes, etc.) may temporarily and partially reduce some stressors, but critics of urban living argued that regular retreats to the solitude and solace of nature provided the only true remedy. Nature, they contended, provides the deep relaxation that restores bodies and minds from the stresses of urban living. With this logic as justification, urban parks and green spaces became popular public health issues in the late 1800s and continue to be so to this day. Park advocates insist that exercise of the mind, body, and spirit in nature cures emotional and physical ailments caused by urban life.

Urban parks of early industrial cities, if they existed at all, were small, ornamental patches of land located near busy crossroads. Their designs intentionally showcased fountains, monuments, and vendors rather than screening visitors from the hustle and bustle of city life. Workers toiled ten to twelve hours a day servicing the Industrial Revolution, lived in cramped conditions, and had no time or opportunity to escape the grime and stress of urban conditions with expensive trips to the ever-receding countryside.

Social reformers directed public attention to what they perceived as a brewing public health crisis. The president of Harvard cautioned in 1914: "The evils which attend the growth of modern cities and the factory systems are too great for the human to endure." The lead architect of Pennsylvania's capital argued forcefully, in 1901, that the city needed a park system to provide "resting places where nervous and tired people, the sick, and mothers with children can secure a complete change of scene, freedom from noise, boisterous games, and the constant passing of people."[11]

Central Park in New York City is one of the first city parks designed specifically with the intention of promoting public health. It was commissioned in 1857 and eventually designed by famed landscape architect Frederick Law Olmsted Sr., who was a staunch advocate of the public health benefits of nature and recreation.[12] Smooth paths winding gently down mild slopes entice strolls that require little mental energy to navigate. Scattered trees and vegetation provide glimpses of sunlit meadows, creating tempting destinations and reasons to linger. Quaking leaves, babbling brooks, and colorful vegetation distract attention away from inner worries about work, finances, and survival. Larger vegetation and sloping hills hide from view and screen from hearing the sights and sounds of the city. Most every aspect of the park's design intentionally facilitates escape from the city into an oasis of nature.

Some park-health advocates believed physical and emotional health resulted not just from leisurely strolls through pastoral landscapes but also through active and unstructured out-of-doors play. The first sand gardens appeared in Boston in the 1880s, justified by the public health concern of promoting physical activity in city children, who otherwise could only shoot marbles in dank alleys. One of the first sanctioned playgrounds was established in 1893 in Chicago. It was staffed by a kindergarten teacher and a police officer. Cities around the country followed these precedents; by

1899, 13 U.S. cities had supervised play areas, 38 cities by 1906, and 504 cities by 1917.

Whether or not the general public believed parks provide public health benefits remains unclear, but the growing parks movement continued to use health as a justification. Very few U.S. cities had any parkland prior to 1890s, but by the 1920s, tens of thousands of urban acres were designated and managed as parks across the nation. Some of the most aggressive parks programs occurred in Los Angeles, Minneapolis, Cleveland, Chicago, Kansas City, Boston, and New York. Parks are now common, if not expected fixtures of all residential neighborhoods. Most twenty-first-century U.S. communities have local ordinances requiring new residential developments to install parks or playgrounds for every fifty or so houses. In addition, most states set minimum standards for community-based recreation opportunities such as hiking, swimming, and basketball.[13]

Efforts to provide children contact with nature extend beyond merely creating open spaces and parks. In the late 1800s and early 1900s, "fresh air charities" organized day trips to rural and natural settings, where fresh air and open spaces would infuse children with wonder and health. Urban schools inserted nature study into their curricula in efforts to compensate for the loss of "natural" activities such as hunting, climbing, and exploring, as well as the lack of environmental knowledge about plant names and natural processes. Children lacking these opportunities were thought to be physically and emotionally damaged. Educators feared that lack of contact with nature produced "hoodlumism, juvenile crime, and secret vice." Books such as How to Study Nature in Elementary Schools, published in 1900, became popular school texts. Summer nature camps, outdoor schools, scouting groups, and after-school programs erupted onto the scene to meet this pressing social need.

Health-related concerns also helped justify the great system of national parks that now protect Old Faithful, the Grand Canyon, and Skyline Drive. For nearly five years, from 1910 through 1915, drafts of the National Park Service's Organic Act included a mission statement with explicit references to "public health." This reference was removed from the final wording, but not the intent, of the 1916 law establishing the National Park Service. Public health benefits of parks were clearly in the minds (and correspondences) of the founders of the U.S. federal park system.[14]

The promotion of public health through parks and recreation is now thoroughly embedded in our institutions. Scientific studies document how nature-based recreation improves family bonding, self-confidence, and physical fitness. Leisure and community recreation programs have been shown to reduce crime, keep teenagers in school, and substitute for drug abuse and gangs. Today the National Recreation and Park Association accredits four-year undergraduate degrees in parks, community, organizational, and therapeutic recreation—all of which address health-related issues. These recreation professionals are employed by public agencies, businesses, nursing homes, and schools to promote public health using recreational activities ranging from volleyball leagues and yoga classes to nature study and summer camps.[15]

Health is notoriously difficult to define in humans, just as it is in ecosystems, so we typically focus on risk, which is easier to measure—risk of a specific disease, injury, or illness. We live in a "risk society," where we balance risk with affluence, freedom, and equity.[16] For example, each of us seems required to bear a toxic burden. Our lifestyles depend upon paper, plastics, petroleum, pesticides, and industrial catalysts whose production results in tons of health-toxic wastes, perhaps as much as a ton per person per year. We ask scientists to estimate the risks of exposure to these toxins, and we ask regulators to allocate risks in an acceptable and equitable manner. The most toxic chemicals are produced by large chemical, metal, and petroleum industries. These industries are highly concentrated and heavily regulated so their pollution may be less risky than the toxins produced by dry cleaning, automobile repairs, film processing, and water treatment, which are highly dispersed, less easily regulated, and closer to residential areas. The U.S. Environmental Protection Agency lists more than 80,000 synthetic chemicals currently in use in North America. About 3,500 of these have been studied sufficiently to estimate their risks to human health, and about 15 percent of these 3,500 are identified as presenting significant health risks, either because of their extreme toxicity or the high levels with which they are present in environments.[17]

How much toxin is acceptable? How much risk? Should we remove toxins up to the point that only one human death is likely out of 100,000, or should we continue removing toxins until risk is reduced to one in a million or one in a billion? The cost of reducing risk to one in a billion might require outlays of time and money so enormous that we must significantly curtail actions that improve human health and well-being in other ways.

Take as an example a community faced with cleaning up carcinogenic PCBs currently contaminating their drinking water. The source of pollution might be the insulation in an old electricity transformer buried in a landfill. After the transformer has been found and removed, it still might cost millions of dollars to reduce concentration levels of PCB in the water supply to a point that only one in a billion people will become sick and die. Perhaps hundreds of lives could be saved if the community instead spent that same amount of money installing smoke detectors in homes and placing intensive care units near where people need them, such as where gunshots and traffic accidents occur regularly. Perhaps thousands of lives could be saved if that money were instead spent improving automobile safety (i.e., better guardrails) or distributing condoms and clean needles that minimize transmission of disease. We don't have the resources or political will to eliminate all risks, so we must constantly make trade-offs.

The facts and figures needed to make these trade-offs are incredibly complicated and require specialized knowledge. Most of us don't possess the time, know-how, or inclination to make informed decisions about these matters. We ask government regulators looking out for the public good to identify risks and decide when these risks become unacceptable; thus we ask a lot of our regulatory system and give it considerable control over how environmental quality is defined and managed.

Despite considerable scientific effort, most risk estimates reflect only tentative and partial understanding of the impacts that chemicals have on human health. There are just too many potential impacts to assess. For example, potential impacts to consider when licensing a new pesticide might include cancer, immune system deficiencies, kidney failure, weight loss, balding, sexual impotence, intelligence loss, or any number of other maladies—each requiring their own expensive testing regimes. Another reason for uncertainty is that multiple modes of contact are possible (i.e., skin absorption, ingestion, or inhalation), each with a different mechanism of impacting health. Moreover, it is probably impossible to test for all possible interactions among all existing pesticides. Each potential impact and each mode of contact may require different tests. Another reason for uncertainty is that we don't experiment on humans and thus must make inferences from animal tests and computer simulations.

We also need to consider the health risks caused by not approving the pesticide. The pesticide in question might increase production of inexpensive foods, reducing malnutrition in people who could not otherwise afford higher-priced foods. There also might be risks inherent in the pest-control methods currently practiced and potentially replaced by the new pesticide. Perhaps the new pesticide will decay sooner and reduce the longer-lasting toxic residue left by the current pesticide. It also may be easier to apply safely, improving the health of farmworkers. Fully assessing the range of risks associated with any new product or procedure, therefore, is just too expensive, assuming we had the science and technology to do so.

To illustrate the difficulty of assessing and deciding acceptable risks, let us focus on just the health risks created by a person ingesting a hypothetical pesticide. We will ignore other modes of contact and other risk factors. Even this simplified task is devilishly difficult. The toxicity of the pesticide to humans must be determined using mice, rats, and pigs. Laboratory animals are administered very high dosages of the pesticide and then observed for several weeks. If malignant cancer appears, then another regulatory process begins (not discussed here) and the chemical is likely banned.

If the animals develop benign tumors, reproductive disorders, depressed immune systems, behavioral problems, or some other symptoms deemed unhealthy, then the task of the health scientists becomes identifying a safe exposure level. Small groups of animals are fed different amounts of the chemical and their symptoms observed. A dose-response curve is calculated that correlates the severity of symptoms with dosage levels. Typically, as dosages decline so do negative symptoms. Statistical models are used to estimate the dosage that would produce no observed effect. The acceptable daily exposure for humans is then set by dividing the no-observed-effect dosage by a safety factor, usually 100, to account for any differences there may be between the laboratory animals' responses and human responses. For example, the dose-response curve may predict that no mice will develop negative symptoms when exposed to less than 1 part in a 100,000. The approved human exposure would then be 1 part in 10 million.

These safe dosage levels obviously contain some uncertainty. But yet more uncertainty exists. Regulators must estimate how the pesticide will be applied, how fast it will decay, how it will accumulate in the food chain, and how much of it a typical consumer is likely to ingest each year. Using these estimates, regulators devise regulations for how much of the pesticide can be applied and under what conditions.[18]

Estimating the health risks of this hypothetical pesticide seem difficult enough; deciding on *acceptable* risk of an existing pollutant is even harder. Take mercury as an example. Mercury is a by-product of burning coal to produce abundant and inexpensive electricity. It gets into the air we breathe and is absorbed by the fish and vegetation we eat. What is an acceptable number of people that can be made sick and/or die from mercury poisoning? Even a small risk of death or problem pregnancies, say one in a million, affects many people in a world of many billions. These risks must be balanced with the risks of eliminating the mercury from coal emissions. What is the added cost to electricity we are willing to accept if smokestack scrubbers or different fuels are required? What jobs will be lost because of higher energy costs? What will be the health consequences of these lost jobs and lower disposable income?

Regulators attempt to quantify and trade off benefits with costs in trying to find answers to these difficult questions. They must estimate the benefits produced by reducing the pollution and compare them with the costs of new pollution-control mechanisms. Such calculations, for obvious reasons, generate considerable debate. Both the costs and the benefits are extremely difficult to calculate and compare because it requires pricing priceless qualities such as a human life and intelligence. The following chapter on equitable nature describes some of these methods and their challenges.

CONCLUSION

People living in developed nations, especially the United States, enjoy, by world standards, an extravagant lifestyle. Dramatic and extensive manipulations of nature are required to provide the comforts and material possessions we enjoy and to absorb the wastes we produce. Our ecological footprints are enormous: requiring tens if not hundreds of acres to provide the resources we use and to dispose of the wastes we generate. While we are not entirely ignorant of the environmental consequences of our lifestyles, we tend not to dwell on them in our everyday decisions. Instead, we take steps to limit risks by developing environmental polices and regulations that curtail our most dangerous excesses and eliminate the most dangerous risks.

One of the early legislative actions clearly motivated by the link between environment and health were the first versions of the Clean Water Act (1948) and the Clean Air Act (1955). A few years later, in 1957, Congress passed the Delaney Amendment prohibiting the use in food production of any substances linked to cancer in laboratory animals. Other acts followed in quick succession, such as the Solid Waste Disposal Act in 1965, the Toxic Substances Control Act in 1976, and the Superfund Act of 1980, to name just a few. The Occupational Safety and Health Act became law in 1970 in direct response to health concerns about modern industrial systems, especially chemicals used during industrial processes. Also in 1970, the authority for pesticide regulation was transferred from the U.S. Department of Agriculture, understandably focused on increasing agricultural productivity, to the Environmental Protection Agency, more directly focused on protecting public health.

Most of these acts and agencies have been amended or reconfigured numerous times. Many laws and regulations explicitly focus on protecting human health from the toxins and environmental changes produced as by-products of our consumptive lifestyles. In some ways, Carson got her wish. While protection against toxins is not in the Bill of Rights, it is firmly entrenched in law and government bureaucracies. But difficult questions remain in our efforts to understand and manage nature: Do our protections and regulations go too far or not far enough? What qualities of the environment are we trying to protect or avoid? What risks will we accept? Which natures do we want?

* 8 *
Fair Nature

Pricing a Life • Industry Relocation
Racism • Denying Development Opportunity
Intergenerational Equity

Any discussion of nature must ask: Whose nature? Who gets to live or work near pollution? Who gets to hike in the wilderness? Who knows enough to engage the political processes that affect land uses near their backyards?

Equity implications are embedded in every decision we make, from product purchases to land-use zoning. Every product we purchase requires resources and produces wastes; someone lives, works, or plays near where these resources are extracted and wastes disposed. Every land-use decision creates opportunities for some people and denies them for others. Even seemingly benign acts of preservation can create inequities because extractive industries may be displaced to communities ill equipped to monitor and manage environmental impacts. Whenever we look at nature, we should see issues of fairness: someone is benefiting and someone is losing.[1]

PRICING A LIFE

In 1991 the vice president and chief economist for the World Bank, Lawrence Summers, wrote a memo that was leaked for wider distribution and critique because of its controversial recommendations to increase pollution in the least developed countries (LDCs).[2] He began the memo by asking, "Shouldn't the World Bank be encouraging MORE migration of the dirty industries to the LDCs?" He went on to point out that the primary costs of pollution come from the lost earnings and lower productivity caused by impaired worker health: "Health-impairing pollution should be done in the country with the lowest cost, which will be the country with the lowest wages. I think the economic logic behind dumping a load of toxic waste in the lowest wage country is impeccable and we should face up to that."

He also argued that lower average life expectancies further justifies increased pollution levels in LDCs because fewer people will live long enough to suffer ill effects from the pollution: "The concern over an agent that causes a one in a million change in the odds of prostate cancer is obviously going to be much higher in a country where people survive to get prostate cancer than in a country where under 5 mortality is 200 per thousand."

Summers later became secretary of the U.S. Treasury under President Clinton and then president of Harvard. It remains debated whether the memo was intended as a serious justification for policy or as an attention-grabbing device to question economic justifications for policy. Regardless, the memo's conclusions reveal how economic calculus and environmental pollution can combine to produce social inequities.

We knowingly distribute pollution. Where it goes is often decided using economic criteria. The World Bank memo brazenly presents economic rationale for polluting places and endangering people where wages are low, lives are short, and people are so focused on subsistence that they willingly sacrifice long-term health concerns. The cold harsh calculus of economics puts a price on human life, trades off human health with industrial profit, and balances human dignity with the cost of doing business. The same sorts of decision criteria affect U.S. environmental policy.

The U.S. Environmental Protection Agency (EPA) is charged with protecting Americans from ourselves. The goods and services supporting our comfortable lifestyles unfortunately produce lethal toxins and damage the ecological systems on which our culture, economy, and health depend. The EPA helps Americans balance our personal pursuits of prosperity, status, and convenience with our social obligations to sustain thriving and healthy communities. Every community's challenge is to find the right mix of economic growth and environmental quality: Unrestrained growth risks depleting resources before substitutes can be found and risks polluting the environment faster than toxicity can be determined and corrected. Slowing growth, on the other hand, risks delaying innovations that save lives and spread prosperity.

Air quality provides an example of the difficult trades-offs we must balance. How clean should your air be? On a typical day, you breathe 3,400 gallons of air. During that typical day, American automobiles, industry, and power generators discharge over 800 million pounds of materials into your air. Some of that material harms your health and degrades environmental quality. As a national policy, we choose to allow these air pollutants because eliminating them unacceptably limits other things we value, such as automobile use, air-conditioned comfort, and economic growth. Many of us refuse to pay higher prices for renewable energy, efficient technology, and recycled products. By doing so, we accept increased health risks and decreased environmental quality as compensation for cheaper utilities and ample consumables. The EPA helps us understand these trade-offs by documenting the costs and benefits of air (and other types of) pollution.

A retrospective analysis of the EPA's clean air programs between 1970 and 1990 shows them to be a pretty good deal for U.S. taxpayers: the costs were approximately $500 billion while the benefits were approximately $22 trillion. Had the results been reversed, the agency would come under pressure to revise its clean air standards and programs. Even though the agency is statutorily prohibited from considering costs in setting standards, it has only finite resources and must decide which programs to fund and which to eliminate. The task becomes all the more challenging when the EPA compares the benefits and costs of its programs with those of other government agencies in competition for citizen support and scarce tax revenue. Regrettably, we have insufficient resources to solve all our problems. Americans must choose among programs that promote safer roads, better health care, higher education, stronger defense, less pollution, and healthier environments.

The devil of these comparisons, of course, is in the details. Costs are difficult to estimate and include operating expenses for government programs, economic opportunities lost due to inflexible regulations, and increased prices for consumer goods and services. Benefits are even more difficult to evaluate because improved human health is among the benefits of these programs. In order to compare the costs and benefits of its programs, the EPA feels pressured to put everything on the same scale: dollars. It therefore prices a human life and multiplies that price by the number of people that would die if air pollution were not reduced. That estimated health benefit is then added to other benefits of reduced pollution, such as the increased timber production and improved tourist economy also resulting from less pollution. These total benefits are then compared to the total cost of pollution control in the benefit-cost analyses used to evaluate EPA programs.

By pricing a life, the EPA can directly compare the value of a life to the cost of pollution controls. An awkward consequence of this approach is that it allows us to trade off human life for economic considerations: the inalienable right promised to U.S. citizens in the Declaration of Independence—the right to life—can be exchanged for economic development. Justifying policy decisions by trading off the value of our lives is not limited to the EPA or pollution control. A variety of agencies and industries use it to inform decisions about everything from setting speed limits to approving new pesticides to designing automobiles.

Various methods are used to calculate the value of a human life. These methods admit the difficulty of valuing someone you know who has children, an active role in your community, and a unique personality. So instead of valuing specific individuals, the EPA values a generic, average, typical "statistical life." One of the most common measures is based on labor markets where employees willingly accept higher wages to work riskier jobs, such as in mining coal. Underground workers have higher health risks than surface workers and, accordingly, get paid more. Aggregation of the pay differential is assumed to represent the price workers place on their lives. Other methods equate the value of a life to the earnings of a lifetime—you are what you earn. Variants of this method reduce the value of a life by subtracting lifetime purchases from lifetime earnings—apparently assuming that misers lead more valuable lives. Still other methods aggregate what people actually pay to minimize risk in their lives (safety belts, smoke detectors, health insurance, and so on). And still other methods examine court settlements in wrongful death lawsuits where juries award monetary compensation for the anguish of losing a loved one. Published estimates range from a hundred thousand dollars to tens of millions of dollars per human life. Estimates the EPA uses in some of its analyses range from less than $1 million to more than $9 million. (As an aside, the EPA also prices human intelligence. Each IQ point degraded by lead exposure, for example, is worth several thousand dollars.)

INDUSTRY (RE)LOCATION

Environmental regulations that enhance environmental quality and protect human health in one region can have the unfair effect of degrading environmental quality and harming human health in another region. Polluters can relocate rather than comply with costly environmental regulations. For a wide variety of reasons, most large U.S. industrial firms now have overseas manufacturing facilities. Some of the relocation patterns appear to reflect companies migrating away from strong environmental regulations toward less stringent localities, often in developing countries in the Southern Hemisphere. In the 1960s and 1970s, some of these countries became known as pollution havens.

These practices are now better regulated, and host communities are no longer ignorant of the environmental costs associated with industry relocation. Moreover, not all polluting industries relocated; many responded to tougher environmental regulations by finding alternative and cleaner manufacturing processes. The industries without financial resources to invest in new technologies or for which alternative technologies did not exist seemed more likely to relocate to pollution havens. For example, industries using highly toxic substances such as asbestos, arsenic, mercury, and benzidine relocated in the 1970s to Mexico, Brazil, and India.

Other industries relocate not to escape environmental regulations in the United States but because laws in developing countries require local processing of locally harvested materials in order to build local economies. That is, these countries no longer allow exportation of their raw resources. They understandably insist that processing be done locally so that local communities benefit from the profit and skill of value-added manufacturing. Other industries relocate in search of better access to supplies, abundant labor, and lower taxes. Whatever the reason behind the relocation, the environmental consequences of the industry get relocated also.

Today's communities constantly struggle to find opportunities for economic development and must balance these needs with the environmental degradations that new developments cause. In 2003 a natural gas–fired electricity generation plant was built in Mexicali, Mexico, just across the California border. The electricity it generated went north to power California homes and industry. California has some of the strongest environmental regulations in America, and the plant features pollution-control technologies that meet California's air-emissions standards. Still, critics contend that the parent company, a California utility, located the plant three miles across the international border in order to avoid the expense of complying with stiffer environmental regulations. The Mexicali location has many advantages for the power production plant. It puts it closer to the resources it needs, near an abundant workforce, and in a community that appreciates the jobs and taxes the plant will generate. The extent to which environmental regulations influenced the location remains debated.[3]

The disposal of toxic wastes illustrates the difficult decisions communities face when trading off economic development with human and environmental health. In 1988 Kassa Island, off the coast of western Africa, accepted money for 15,000 tons of ash from Philadelphia's municipal waste incinerators. The ash likely contained heavy metals and other toxic chemicals. It was transported by a Norwegian ship, dumped, and labeled "raw material for bricks." Reports from the dump site in Kassa Island complained of serious environmental degradation and increased health problems of local residents. Threats of legal actions and increased political pressure forced the transport company to remove the material and to change its plans to dump another 70,000 tons of ash that Philadelphians did not want to bury in their backyards.

This is not an isolated example. Many U.S. municipalities ship their wastes, especially their toxic wastes, away from their residents. Waste trade is a multibillion-dollar business. In the 1980s a cargo of waste crossed a national border almost every 5 seconds, 24 hours a day, 365 days a year. As awareness of the trade in toxic wastes grew, many third world countries pressed the United Nations for an absolute ban on waste exports, calling such exports "toxic terrorism." Industrial nations rebuffed these proposals. The Basel Convention of 1989 crafted an international treaty that requires waste exporters to notify and receive permission from waste importers before making shipments. U.S. opposition doomed a treaty amendment that would keep nations from shipping wastes to countries that had standards for waste disposal lower than their own. However, some countries working unilaterally, many in the European Union, have agreed to policies that ban exporting wastes to nations that cannot process them safely.[4]

Toxic waste disposal provides extreme examples of environmental injustice but is really just one of the fairness and equity issues linked to everyday land-use decisions. Since the average U.S. citizen consumes much more than people of other countries, we bear a disproportionate blame for these environmental injustices. Hazards associated with production are incurred by the labor force proximate to the industry: pollution and other by-products stay where manufacturing occurs while the product and profit may be exported. Consequently, wealthy consumers of economic goods rarely suffer the environmental consequences of their purchasing behaviors.[5]

ENVIRONMENTAL RACISM

Environmental quality, like economic opportunity, is not evenly distributed in the United States. Society's least powerful members bear a disproportionate exposure to landfills, strip mines, heavy industry, abandoned lots, and eroded soils. These same people typically enjoy fewer open spaces, parks, vistas, and other environmental amenities. NIMBY (not in my backyard) efforts export unwanted environmental conditions from the backyards of the politically powerful to the backyards of the less powerful. Factories, dumps, highways, and other "developments" typically are not located near people with the resources to oppose them, while desired parks and open spaces are. Political power determines who wins and loses land-use battles.

In 1988 the largest toxic waste dump west of Alabama was located near Kettleman City, in Kings County, California. The local community was blissfully ignorant of the dump's location and operations, even though operators had been charged with million-dollar fines for over fifteen hundred environmental violations at the site. Ignorance shifted to outrage in 1988 when the dump operators proposed adding a facility to burn over 100,000 tons of toxic wastes every year. Not only would the incinerator degrade local air quality, but it would dramatically increase the wastes being trucked through the community. Local residents were mostly Latino farmworkers, roughly 40 percent of whom spoke only Spanish. The county, which was 65 percent white, organized the permitting process. It already earned $7 million per year, about 8 percent of county revenue, from the fees and taxes paid by the waste disposal industry and stood to double that with revenue from the incinerator. The county produced a 1,000-page report documenting acceptable environmental impacts, which it refused to translate into Spanish. The law required a public hearing, so one was scheduled forty miles away from the site at the county fairgrounds in Hanford, the county seat. About two hundred people came by bus and carpool from Kettleman City to testify at the hearing.

The meeting seemed designed to intimidate rather than facilitate public dialogue. The planning commission—all white, all living in a different part of the county—sat behind a table on a raised platform looking down on fifty rows of seats. Behind the seats were bleachers for overflow seating. At the very rear of the auditorium, well behind the bleachers, a hundred yards away from the planning commission and the microphones for testimony, sat twenty-five empty chairs circled around one lone translator. The Kettleman residents had brought their own translator and wanted to testify directly to the planning commission. "That request has been denied," they were told, "the translation is taking place in the back of the room and it won't happen here." The racist parallel to forcing African Americans to sit at the back of the bus was hard to ignore. The residents revolted, disobeyed the planning commission, and testified from the front of the room. That hearing ended the public's ability to comment and the permit was granted.

Residents, aided by Greenpeace and the citizen group El Pueblo, filed a lawsuit to deny the permit. The court agreed that residents of Kettleman City had not been meaningfully included in the permitting process. The county decided not to appeal or to redo the process. The company did appeal but withdrew it application in 1993 in response to increasing attention by local and national press and an intense letter-writing campaign targeting the local Board of Supervisors and Farm Bureau.[6]

INEQUITIES OF PRESERVATION

Land decisions to preserve nature and create national parks are anything but benign. While nature gets protected and some people enjoy enhanced recreation opportunities, other people get demeaned and displaced.

The Great Smoky Mountains National Park was dedicated in 1940 after nearly two decades of intense campaigning by and for urban middle- and upper-class interests over the interests of local people. It had the laudable goal of creating a large charismatic park that would be accessible to people living in large urban areas along the East Coast. To achieve this goal, promoters first portrayed the area as empty of civilization and a soon-to-be victim of rapacious capitalism that would strip away the abundant economic resources, leaving behind a barren wasteland. The promotional campaign attacked industry and greed.

> Why should future generations be robbed of all chance to see with their own eyes what a real forest, a real wildwood, a real unimproved work of God, is like....
> There is no use, then, in talking about conserving the Smoky forest by turning it into a national forest after the lumbermen get through with it. The question, the only question, is: Shall the Smoky Mountains be made a national park or a desert?[7]

Industry, however, turned out not to offer much resistance to the park's creation when John D. Rockefeller used $5 million of his personal fortune to purchase land from timber companies and other economic interests. The real obstacle to the dreams of park promoters became the mountain residents who did not want to be told that the thousands of acres surrounding their communities could no longer be used to provide jobs, generate taxes, and sustain their culture. The local people opposed abandoning their land or lifestyle in order to satisfy the aesthetic interests of urban elite.

Park promoters began a public relations campaign to build public support among the politically powerful urban population in order to overcome local rural resistance. First, they promoted the social and economic benefits of having a park near urban populations. They sold tourism to local business interests, and they sold spiritual rejuvenation to the weary masses looking for escape from confined urban living. Second, they promoted an image of the area as empty, and in the scattered locations where human habitation occurred, they promoted an image of the inhabitants as uncivilized, unkempt, and uneducated hillbillies.

The mountain residents had little political power but organized at least two petitions arguing against the park's creation. All the major newspapers in the region, however, supported the park and little came of these petitions. Promoters characterized local people as cultural primitives who were isolated and socially undesirable. They were portrayed as lazy people, living a subsistence lifestyle, shunning hard work, lacking industry, and concerned only about feuding and bootlegging. They supposedly lacked the capacity to feel love, loyalty, and responsibility. Instead, they tended toward violent and immoral behavior. The park was presented as an opportunity to improve and modernize this backward culture. Some promoters also tried to characterize the mountain people as rustic and quaint. They argued that mountain culture could provide a tourist attraction and curiosity. Roadside displays of hillbilly crafts, music, and culture would provide amusing contrast to sophisticated urban visitors. This aggressive and eventually successful public relations campaign created a negative impression of Appalachian culture, which lingers to this day.

With establishment of the Great Smoky Mountains National Park, local people lost their land and livelihood, and many relocated. A culture was destroyed. This campaign to preserve nature and create a national park demonstrates how all land uses have equity implications. When looking at the wild nature in the Smoky Mountains, we should see not just wildness and biodiversity, but a place where the interests of some people, namely, urban tourists and ecologists, replaced the interests of other people.[8]

These issues are particularly relevant in developing countries. In the urge to protect nature, some well-intentioned, well-off people donate money to purchase tracts of rain forests or other ecosystems. These practices can remove vast land areas from economic development, evict local residents, and limit opportunities for social advancement through harvesting of natural resources. The American economy was constructed several hundred years ago by exploiting local resources: forests were denuded, rivers dammed, and soils eroded. This exploitation fueled the economy that created the lifestyles we now enjoy. It also motivated the conservation policies that now protect these resources. Enforcing preservation of land in other countries denies others the same choices we had and reduces their opportunities to control their future. These trade-offs between social development and nature preservation may be justified because of larger worldwide concerns about biodiversity and carbon sequestration. The only point being made here is that these equity concerns should be explicitly considered when advancing land-use policies.

Clearly, nature preservation can impose a socially conservative political agenda. It protects those comfortable with the current distribution of risks, opportunities, and environmental conditions. Those feeling repressed and hopeless might voluntarily take a chance on change. For example, the loss of biodiversity caused by clearing rain forests for income and food may be unacceptable to those currently well fed and with employment not dependent upon agricultural profit. But people who are undernourished and without an economy might be more willing to accept these losses. People without ample income, job security, and dignity might prefer the income flow from active environmental management over the preservation of wild conditions.[9]

INTERGENERATIONAL FAIRNESS

Of course, equity questions don't stop with concerns about those of us alive today: we must look ahead and to our obligations to future generations. Defining these obligations is not easy, and one way we talk about them is in the context of "sustainability."

Sustainability has many definitions. The contrast between so-called weak sustainability and strong sustainability provides one way of exploring our obligations to future generations.[10] Policies consistent with weak sustainability strive to sustain or increase the social and natural capital needed by future generations. Each generation is obligated to provide the next generation with a combination of wealth, technology, knowledge, ecological services, and other social and environmental qualities that meet or exceed what was inherited from their parents. Weak sustainability allows some increases in wealth and technology to substitute for some decreases in ecological services and environmental qualities. For example, the current generation can satisfy obligations of weak sustainability even if they irreparably pollute watersheds, as long as they also provide future generations the wealth and knowledge to provide for water-purification technologies. What matters is that future generations have sustainable flows of desired goods and services (water, oxygen, wood, recreation, etc.); it does not matter how the flows are sustained. The popular Brundtland definition of sustainability has been characterized as advocating weak sustainability by promoting "development that meets the needs of the present without compromising the ability of future generations to meet their own needs."

The requirements of strong sustainability, in contrast, would not allow us to substitute social capital for natural capital. Strong sustainability requires that future generations inherent functioning ecological systems, not just sustainable flows of specific goods and services. Wealth and technology, it is argued, do not provide equivalent substitutes for ecological or cultural opportunities. Land uses that generate wealth by destroying natural history, biodiversity, or ecological functions are inappropriate. Destroying these qualities removes or restricts opportunities for future generations. For example, municipal water filtration and chlorination plants might not substitute for functioning ecological systems that clean and store groundwater because the dependency on technologies restricts choices by forcing future generations to sustain that technology.

The Pacific Northwest provides a classic example of the lost, restricted, and created opportunities that make deciding between weak and strong sustainability so difficult. During the 1900s old-growth forests, salmon fisheries, hydroelectric power, and other natural resources were rapidly converted into a regional infrastructure of wealth, universities, government, roads, cities, power lines, high-tech industries, and industrial agriculture. Criteria for weak sustainability would be met if future generations have the same or better opportunities for living, and living well, using this new regional infrastructure. For example, the next generation of people likely will find successful careers writing software, own comfortable suburban homes, and have safe drinking water. Criteria for strong sustainability might not be satisfied because energy, clean water, and even salmon would need to be imported or manufactured. Also, future generations might have fewer opportunities to live in communities defined by cultural traditions associated with managing, harvesting, and processing natural resources such as old-growth forests and swarming salmon fisheries. Future generations will be restricted in their choice of cultural and environmental opportunities because of choices and changes previous generations made. However, future generations also will have new opportunities because of the choices and changes previous generations made. Which opportunities do we want to sustain and pass on?

One answer to this question argues that it does not matter what opportunities we create and leave behind because the unborn are not part of our moral community; we have no social contract with them and therefore no obligations to them.[11] Obligations in contemporary society are based on a social contract of reciprocity, and because future generations can do nothing for us, we need not do anything for them.

The counterargument suggests that we do have obligations to future generations, they are part of our moral community, and they do, in fact, do something for us. Future generations carry on the traditions and ideals that we value. We owe them something for honoring our memories. We fought wars to protect freedom and democracy; certainly we feel future generations should and will continue to value these ideals. For this reason alone, future people are members in our moral community. Of course, we also will have familial bonds with people in the future; hence they belong to our moral community as family members. According to this reasoning, we owe these people democracy, literacy, and opportunities to pursue happiness. Perhaps we also owe them certain environmental qualities.

Another argument against future obligations is that we can't predict what will be important then. If we look back several hundred years to the 1700s, it is unlikely that anyone could have predicted our current needs for fossil fuels, Internet privacy, or wilderness preservation. Perhaps we are just as unlikely to know what future generations will need. As a result of our ignorance, the argument continues, we can't possibly know what opportunities to sustain. However, the counterargument suggests that *all* humans—past, present, and future—will desire a healthy diet, clean air and water, ozone protection from radiation, and various sources of pleasure. Perhaps we should sustain these basic life-support environmental qualities. We can also look back to the 1700s and see that the founding fathers were right in their educated guesses that freedom, equality, and justice continue to be valued today. We can expect them to be valued in the future. Perhaps we also should make educated guesses about the environmental qualities that will be valued in the future and manage our environment accordingly. Some environmental qualities—such as the frontier, agrarianism, wilderness, and biodiversity—contribute significantly to the American culture. The only reason our founding fathers did not discuss them in the Constitution and Bill of Rights, this counterargument continues, is because they could not envision the magnitude of environmental change that would be made possible by modern technology and economy.

Choosing among opportunities to leave the future is no easy task. We can't even decide if doing so is a good idea. If we do decide to do so, then we must recognize that the opportunities we consider will reflect our assumptions about nature and the future. If we conceive of nature as robust and human technology limitless, as was the case when Europeans were settling the Americas, then we might want to dominate nature and build farms, factories, infrastructure, and technological developments that provide control over the miseries nature can cause. Future generations may appreciate established universities to educate their young, automated factories to produce their goods, biotech agriculture to satisfy their hunger, convenient energy to power their homes, progressive governance to protect their rights, roads that take them to jobs and tourist destinations, and medicine to cure their ills. If, on the other hand, we conceive of nature as fragile and human ingenuity limited, then we may want to minimize our environmental degradations so that future generations have a habitable planet with resources and opportunities to meet basic needs.

CONCLUSION

Environmental management can be fair or unfair in how it distributes opportunities and pollution. Not surprisingly, the modern environmental movement, like its late nineteenth-century predecessor, increasingly emphasizes fairness.[12] The modern environmental and civil rights movements are joining forces, intentionally doing so beginning in the early 1980s. Concerns about equity and fairness permeate both movements. The emerging movement, sometimes called environmental justice, may be one of the most powerful and important social movements of generations. It unites concerns about public health, worker safety, job security, hazardous wastes, global economics, land use, transportation, housing, resource allocation, community sustainability, and wilderness preservation. It connects concerns about racism with appreciation of biodiversity. It bridges equity and environment, and it forces us to move beyond the polarizing human-nature dichotomy by its focus on both the human *and* environmental consequences of every action.

Our every decision affects the income, community, and environment of others. As discussed in the next chapter, doing unto others as we would have them do unto us has implications for the types of nature we want to create, preserve, and sustain. As will be discussed in chapter 11 ("Rightful Nature"), extending the moral community further, to nonhumans, makes concerns about equity even more inclusive.

Spiritual Nature

Dominion · *Caring for Creation*
Do unto Others · *Inspiration*
Garden of Eden · *Work Is Divine* · *Apocalypse*

Darwinian evolution can be a bitter pill for the devout. It replaces divine intervention with random events, suggesting that humans are fortuitous accidents rather than inevitable creations. Genesis, at the beginning of the Old Testament, provides an alternative creation story, one that most people prefer. Opinion polls show that 45 percent of Americans believe God created things pretty much as Genesis describes it. Another 37 percent accept that the fossil record indicates millions of years of evolution but believe God guided the process. Only 12 percent believe evolution unfolds as Darwin described.[1]

Many of the world's religions have creation stories, and most also contain land ethics about humanity's responsibility for nature. The Jewish and Christian religious traditions that dominate post-Columbus North America are emphasized in this chapter. These traditions, however, are rarely explicit about environmental ethics. None of the Ten Commandments, for example, speak explicitly about stewardship of creation, and most of the Bible focuses on relationships between God and humanity or on relationships among humans, not on the relationship between humans and nature.[2] As a result, numerous and sometimes contradictory environmental themes have been identified by environmental theologians: (1) subdue and dominate nature for human benefit; (2) care for God's creation; (3) treat others as you want to be treated; (4) experience God through nature; (5) re-create the Garden of Eden; (6) avoid waste and rest, which are evil, in favor of work and industry, which are divine; and (7) absolve yourself of environmental responsibility because divine intervention will soon end everything with the Apocalypse.

Genesis explains humanity's special role on Earth. In well-known passages of the creation story, God gives humans dominion over fish, birds, cattle, and "every living thing that moves upon the earth." Light, night, soil, wind, rain, and fire exist for human use. Other sections of the Bible make it clear that being human means being different than nature and that Earth is but a temporary home—the soul is eternal and transcends Earth. Of all creation, only humans are made in God's image. Jesus, Son of God, was both human and divine. He walked Earth as a man, not as an animal, a plant, or a mineral. Environmentally minded critics say that the Judeo-Christian land ethic invites environmental abuses by suggesting that humans are special and thus more important to God than is the rest of creation.[3]

Related critiques suggest that the Judeo-Christian land ethic facilitates environmental abuses because its conception of time is linear, with a distinct beginning and end. God created Earth, humanity plays out its drama on Earth, and God will end Earth. Humans are not reincarnated; they do not return to Earth. Spirits of ancestors do not dwell on Earth; they dwell with God. Thus, humans do not need to create a pleasant place on Earth for the afterlife because life on Earth is temporary and secondary. Moreover, God promises to sustain and nurture his chosen people, so fears about a finite Earth and resource depletion are misplaced; miracles will feed many people from one loaf of bread. Salvation depends upon preparing the soul for heaven, not conserving resources on Earth.

The very beginning of the Bible contains two creation stories. In the first story, God took six days to make heaven, light, water, land, plants, animals, and finally people (Gen. 1:1–2:4a). God created man and woman in his image and gave them dominion: "Be fruitful and multiply, and fill the earth and *subdue* it; and have *dominion* over the fish of the sea and over the birds of the air and over every living thing that moves upon the earth" (Gen. 1:28).[4] Later in Genesis, God brings all the plants and animals before Adam and allows him to name them. Some interpreters of this text claim that the act of naming confirms humanity's superiority over nature.

"Subdue" and "dominion" are strong words that seem to suggest that God cares more for humans than the rest of creation. However, the words are infrequently used in the Bible and their interpretations are contested.[5] Some Old Testament passages suggest having dominion means playing a pivotal role in the situation, or even serving rather than subjugating the other, as a husband supposedly has dominion over his wife and God has dominion over humans. "Subdue" is less ambiguous. The Hebrew translation is used in ancient Hebrew texts to describe military actions with undertones of power and violence, such as putting under one's feet, conquering, violating, bringing into bondage, and ruling over. Focusing just on these interpretations of "subdue," it seems God commanded Adam to make war on Earth.

While in the Garden of Eden, Adam and Eve were charged by God to live lightly on the land, eating a vegetarian diet: "And God said, Behold, I have given you every herb bearing seed, which is upon the face of all the earth, and every tree, in the which is the fruit of a tree yielding seed; to you it shall be for meat" (Gen. 1:29). Life in the Garden of Eden is idyllic, with lambs and lions lying together in bounty, peace, and innocence; but once God banished Adam and Eve from Eden, things became more difficult. Nature becomes adversarial: " . . . cursed is the ground because of you; in toil you shall eat of it all the days of your life; thorns and thistles it shall bring forth to you; and you shall eat the plants of the field. In the sweat of your face you shall eat bread till you return to the ground, for out of it you were taken; you are dust, and to dust you shall return" (Gen. 3:17-19).

Repeated human sin brought forth God's wrath in the form of a flood, on which floated Noah's Ark and a great deal of biodiversity. After ten long months underwater, mountains peaked through the receding flood. Noah unloaded his charges on the Ararat mountain range, and God provided new instructions about the relationship between humans and nature. Humans would no longer live at peace with the rest of creation; rather, animals would be meat and live in fear of humanity: "The fear of you and the dread of you shall be upon every beast of the earth, and upon every bird of the air, upon everything that creeps on the ground and all the fish of the sea; into your hand they are delivered. Every moving thing that lives shall be food for you; and as I gave you the green plants, I give you everything" (Gen. 9:2-3).

In other biblical stories, God uses nature to test and punish humans. Life on Earth is harsh. It tests a soul's merits for life in the hereafter. Cast from the Garden of Eden, humans must work hard to survive. Nature is neither plentiful nor beneficent; rather, it is cruel and limiting, with dangerous plants and animals. God repeatedly shows his displeasure with pests, famines, floods, and other ecological disasters. Satan, death, and the demise of civilization lie in the untamed wilderness, so God's chosen people must bring faith to the wilderness and maintain faith during the harsh tests of environmental deprivation. The natural world is a "hideous wilderness," a "howling desert," that has "fallen" from grace and required intervention and improvement.[6] Good Christians did God's work by settling, taming, and conquering nature's wildness.

Regardless of hotly debated interpretations of biblical passages, the monotheism of the Judeo-Christian tradition dramatically changed people's understanding of nature. The landscape existing before the God of Abraham was inspirited, alive with many gods, demons, and ancestors. Sacred markers and sacred places abounded and demanded respect. One did not blithely kill, chop, dam, divert, flood, fill, or pave over the home of gods, as is suggested by this Native American spiritual leader's response to requests to adopt European land-management practices.

> You ask me to plow the ground. Shall I take a knife and tear my mother's bosom? You ask me to dig for stone. Shall I dig under her skin for bones? You ask me to cut grass and make hay and sell it, and be rich like white men. But how dare I cut off my mother's hair?[7]

Judeo-Christian teachings de-spirited nature by claiming that nature, unlike people, is not caring, feeling, or inspired. Spirits on Earth exist only in people, and thereafter reside elsewhere, in heaven or hell. With only one God to worship, the spirits embedded in nature were banished, buried, ignored, and forgotten, as were their landscape totems. Earthworks, sacred groves, stone carvings, and other monuments were plowed under, broken up, and left to decay. Christian preachers exorcized the landscape, desacralizing and taming it, with churches replacing natural areas as sacred places.

CARING FOR CREATION

Other biblical texts and Christian and Jewish teachings emphasize caring for creation rather than subduing it. They point out that God made creatures for his sake and theirs, rather than just for humankind alone. In the creation story of Genesis, God judged *all* of his creations "good." Thus, all parts of nature are valued by God and deserve protection. Nature has value beyond how it serves humans. Numerous passages, such as Proverbs 12:10, acknowledge the value of nonhuman life: "A righteous man regardeth the life of his beast." Christian thinking along these lines supported, and some argue started, the campaign against unnecessary cruelty to animals beginning in the sixteenth century.[8]

Some passages in Genesis soften, if not contradict, the interpretation of dominating and making war against nature. The second telling of the creation story begins at Genesis 2:4b. Although it appears second, this version is older and it reflects the oral, arid, and tribal traditions of ancient Hebrews. Creation occurs in a different order, with Adam being formed from dust before God creates trees, rain, Eden, and finally Eve. Rather than dominate and subdue creation, humans are instructed to "till and keep" it (Gen. 2:15). In some translations, "till and keep" are presented as charges to "dress," "serve," "care," and "guard" creation. "Tilling" has been interpreted to mean a loving relationship between husband and wife and between God and humans. "To keep" means to guard or to watch over, thus maintaining the capacity of creation to flourish. Humans are charged, then, with guarding what God entrusts to them—creation. Rather than exploiting nature for personal gain, humans should be servicing God's Earth. Tilling and keeping become not just an obligation but a form of worship.

Capitalism's advocacy of private property rights borders on idolatry. Passages in Genesis and elsewhere make clear that the ownership of creation is never in doubt: humans do not own and will not inherit land. God, the landlord, put humans in charge of managing creation and expects it to be nurtured and sustained. God values his creation and only God knows the best condition of nature. By remaking nature in their image, humans might be committing sins of hubris and pride. Humans, as God's agents, are responsible for keeping creation in a condition that would please God— human salvation might even depend upon it. The landlord will evaluate how well his estate has been managed with humans accountable to God for the quality of their leadership and management. Since reckoning might come at any time, stewardship, a form of worship, must be continuous.[9]

Wendell Berry, a farmer/poet/philosopher/preacher, argues that Judeo-Christian traditions do not promote the dualism that environmentalists say damagingly separates humans from nature, soul from body, spirit from matter, and worship from work. He agrees that a damaging dualism exists, but attributes it to Western thought, not Christianity, and offers Genesis 2:7 as an example: "The Lord God formed man of the dust of the ground, and breathed into his nostrils the breath of life: and man became a living soul." The Western dualistic mind concludes that man = body + soul. Berry, in contrasts, interprets the passage as suggesting soul = dust + breath. God did not make the body as a repository of soul, but rather made soul out of earth and Holy Spirit. God created both the dust and the breath. Both the earth and his spirit are holy and combine to create the divine gift of living souls.

The dualist interpretation separates the soul from the body and spirit from earth, and, consequently, encourages us to focus on differences between nature and humans, ecology and economy, when we should instead focus on how ecology and economy help us steward creation. Dualism is harmful if the dualist soul becomes the focus of stewardship while dust gets neglected, used, or abused. In contrast, if we can overcome dualism to think of ourselves as part of nature and as souls living in the midst of creation, then "everything we make or do cannot help but have an everlasting significance for ourselves, for others, and for the world."[10]

The acts of living—economy, leisure, democracy—can be a form of worship showing respect, appreciation, and reverence to God and creation. Economy and ecology, rather than competing ideals, can become ways of understanding and respecting creation. Rather than polarizing environmental debates around economy and ecology, religion can serve to unite economy and ecology as ways to understand one pole, creation. God, then, becomes the other pole. Religion can help us understand the relationship between the living souls of creation and God. Economics and ecology may have much more in common than they realized: while they have different methods, they can have the same goals.

"God's Greens" is a label given to the growing chorus of voices advocating an environmentally oriented theology and environmentally active congregations. Genesis' charge to care for creation provides just part of the rationale. Readers of the Bible can debate interpretations of dominionism and dualism, but it is harder to debate the Golden Rule. One of the foundational principles of many Judeo-Christian religions is treating your neighbor as you would want your neighbor to treat you. It is also hard to deny that great inequities and injustices result from environmental policies and actions (see chapter 8). The enormous inequities of environmental pollution and exploitation demand religious concern and action. Regardless of whether humans are obligated to care for creation, humans are obligated to care for one another, and doing so has enormous implications for environmental management.

A 1990 papal decree by Pope John Paul II affirms these concerns, stating that environmental degradation damages not just creation, but our human neighbors.

> I. IN OUR DAY, there is a growing awareness that world peace is threatened not only by the arms race, regional conflicts and continued injustices among peoples and nations, but also by a lack of DUE RESPECT FOR NATURE, by the plundering of natural resources and by a progressive decline in the quality of life. The sense of precariousness and insecurity that such a situation engenders is a seedbed for collective selfishness, disregard for others and dishonesty.
>
> . . .
>
> 7. The most profound and serious indication of the moral implications underlying the ecological problem is the lack of RESPECT FOR LIFE evident in many patterns of environmental pollution. Often, the interests of production prevail over concern for the dignity of workers, while economic interests take priority over the good of individuals and even entire peoples. In these cases, pollution or environmental destruction is the result of an unnatural and reductionist vision which at times lead to a genuine contempt for man.[11]

PROOF AND INSPIRATION

Ever since the creation of the world His eternal power and divine nature, invisible though they are, have been understood and seen through the things He has made.
(Romans 1:20)

Much early scientific inquiry had religious purposes. It was a holy pursuit. God was assumed to be the great designer and ultimate watchmaker. Books with titles such as *The Wisdom of God Manifested in the Works of Creation* published in 1691 interpreted nature as evidence of God and as lessons from God. Serious scholars began their studies with the assumption of intelligent and intentional design. The intricate, finely tuned workings of nature seemed too perfect to be the product of random events. Sir Isaac Newton, for example, explicitly interpreted the universal "laws" of force, mass, acceleration, and gravity as evidence of God's logic and wisdom. Thomas Jefferson, upon learning from fossil records that giant woolly mammoths had once walked North America, refused to believe that extinction was possible. Extinction would mean the original design was erroneous in the creation of unnecessary species or unsustainable habitats. He commissioned scientific exploration to find living mammoths.[12]

Proof of intentional design was found everywhere scientists looked. They found it in themselves: incredibly complex, balanced, beautiful, conscious, and creative beings. They found it in the finely tuned fit between animal needs and animal habitats. Hawks, for example, are able to see great distances so they may find food while soaring the skies. These many ecological interactions seemed to miraculously sustain a biosphere perfectly suited for human life. Further proof of divine intervention was found in the rational ordering of day and night: the sun rising in the morning to provide light by which humans conduct business and setting in the evening to make sleeping easier. Even predation was interpreted as intentional and intelligent design. It controlled grazing animals from overpopulating and destroying habitat. Predators merely weeded the herd of old and sick individuals in order to sustain the habitat and preserve the balance of creation.

John Calvin (1509–1564), a contemporary of Martin Luther, is among the most influential Christian theologians. His work is credited not just with reforming Christianity, but initiating a type of thinking that enabled science, capitalism, and democracy. He is a father of modernity. His framing of Protestant theology shaped the thinking of noted environmentalists such as John Muir. Calvin begins his treatise *Institutes of the Christian Religion* with strong and explicit statements that the complexity and nuance of nature prove God's existence:

[God] daily discloses himself in the whole workmanship of the universe. As a consequence, men cannot open their eyes without being compelled to see him. . . . [U]pon [God's] individual works he has engraved unmistakable marks of Glory, so clear and so prominent that even unlettered and stupid folk cannot plead the excuse of ignorance.[13]

Nature not only provides proof to some of God's existence; it also provides access to God's wisdom and a source of inspiration. God, it has been argued, produced two Good Books. One was revealed to saints and prophets and translated into human language to create the Bible. The other book is nature. It was written by God's own finger. Nature is inerrant, unfiltered by language and culture, and accessible to all, regardless of literacy. The Calvinist preoccupation with man's sinfulness promotes distrust of human works. Untouched nature represents unvarnished access to God, hence many Protestants embraced it. Looking to nature for insights about God pervaded American thought in the eighteenth and nineteenth centuries.[14]

Nature is God's grandest cathedral. Outdoor meditation, prayer, and worship enhance religious experiences. In fifteenth-century England, some Protestants began advocating that prayers and services be held in nature as well as in churches. One minister abandoned his indoor pulpit to preach from an elm tree in the church courtyard.[15] Jonathan Edwards (1703-1758), one of America's important and influential theologians, repeatedly found deep religious experiences in wild settings:

> God's excellence, his wisdom, his purity and love, seemed to appear in every thing; in the sun, moon, and stars; in the clouds and blue sky; in the grass, flowers, trees; in the water and all nature. . . . And scarce any thing, among all the works of nature, were so sweet to me as thunder and lighting. . . . I felt God . . . at the first appearance of a thunderstorm. . . .[16]

Transcendentalist icon Ralph Waldo Emerson (1803-1882) also found God in nature: Nature is God; God is nature. Nature is soulful and infused with God's spirit, and creation embodies holy truths. Prophets since Abraham went to the wilderness to find God and inspiration. The proper attitude toward a God-filled nature is wonder, worship, joy, and celebration. Nature is sacred; it conducts divinity. It provides the means to transcend the here and now and to access higher, deeper, more profound truths created by God. Humans fell from grace and were evicted from the Garden of Eden, but the Garden of Eden—nature—remained innocent and uncorrupted. Cities became homes of the banished. Immersion in nature and withdrawal from civilization offered a path toward redemption.

This respect for and worship of nature likely motivated many environmental preservation actions in the United States. Wild places were portrayed as temples to worship rather than as resources to be wisely used for economic gain. John Muir (1838-1914), a pillar of the American environmental movement, found in nature divine inspiration and evidence of God. He preached salvation through nature worship and baptism in wilderness. In a famous passage, he attacked those wanting to use the valley of Hetch Hetchy, located within Yosemite National Park, as a reservoir to serve the water needs of nearby San Francisco. Muir left no doubt that nature was a place where God should be worshiped: "Dam Hetch Hetchy! As well dam for water-tanks the people's cathedrals and churches, for no holier temple has ever been consecrated by the heart of man." Those who pursued the "almighty dollar" over nature worship were doing the "devil's work."[17]

America's influential founding father Thomas Jefferson (1743-1826) also believed in and practiced a type of natural theology. However, in contrast to Muir and Emerson, Jefferson's God perfected but did not reside in nature. Jefferson was a product of a pragmatic, economically motivated southern culture rather than Emerson's religiously inspired Puritan culture. His state was established by the Virginia Company of London with the charge to profit from tobacco trade, not by Puritan zeal to escape religious persecution and create God's country. Nature, for Jefferson, was something to be used and understood through application of rational thought. Studious use of nature brought people closer to God, built moral virtue, and generated a citizenry capable of supporting a democratic government. He did not find spirits or direct revelations in nature. Instead, he found God's truths revealed through the scientific study of nature and through labors directed toward farming and improving the land. As reviewed in chapter 13, "Moral Nature," Jefferson appealed to the moral authority embedded in nature in most of his writings, including the Declaration of Independence.[18]

Spirituality was embedded in much of the nature writing throughout the twentieth century. Popular stories told how nature signified "the smiles of God," "the visible garment of God," the "art of God," the "thoughts of God," and "the playground of the soul." The patriarch of modern nature writers, John Burroughs (1837-1921), observed at the dawn of the twentieth century that "every walk in the woods is a religious rite. If we do not go to church as much as did our fathers, we go to the woods much more, and . . . we now use the word nature very much as our fathers used the word God."[19] To those holding these beliefs and understandings, nature is valued for the spiritual connections it embodies and evokes. Destroying nature is blasphemous. Ignoring nature is ignoring God.

A dramatic example of natural theology comes from the work of one of the more popular nature writers in the early 1900s. In one of his many writings, Ernest Thompson Seton (1860-1946) used nature to justify the Ten Commandments. Observations from his studies of nature proved to his satisfaction that "the Ten Commandments are not arbitrary laws given to man, but are fundamental laws of all creation."[20] In a 1907 paper, he summarized years of observations supporting his certainty in this conclusion. The first four commandments deal with man's relationship to God (have no other God, have no graven images, do not take the Lord's name in vain, keep the

Sabbath holy). He reasoned that since animals do not have a spiritual relationship with a supreme being, it is not surprising that little evidence of these first commandments is found in the behavior of animals. But Seton went on to describe animal behaviors he felt supported the last six commandments, which instruct humans how they should treat one another: "the first four commandments have a purely spiritual bearing; the last six are physical. Man is concerned with all, the animals only with the last six."[21]

Seton explains that the commandment that children shall honor their parents is reflected in the animal kingdom by numerous examples of how animals not heeding their parents die early deaths from starvation or predation. The proscription against murder is illustrated by observations of cannibalism within a species: while cannibalism is observed in the lowest forms of life, it becomes rare and more "repugnant" in the "higher" animals. For the commandment against adultery, Seton argues that adultery is synonymous with impurity, and that animals only breed within their species. He observed that the more promiscuous species are less healthy and less advanced. "The promiscuous animals to-day—the Northwestern rabbit and the voles—are high in the scale of fecundity, low in the scale of general development, and are periodically scourged by epidemic plagues." Seton also observes animals protecting their territory and concludes that animals care about and respect one another's property, thus following the commandment against stealing. And in regards to the commandment against bearing false witness, he uses the example of dogs losing respect and the support of the pack if they signal the pack to follow a false lead in their pursuit of prey. He describes animals being banished or punished for using the "home" of another as proof that animals follow the commandment against coveting a neighbor's house and property. For example, he observed a hen taking possession of another's nest and laying her eggs in it during the owner's brief absence. The result was a confrontation in which all eggs were destroyed and both chickens were losers.[22]

Natural theology (i.e., looking to nature for lessons from God) held sway from the sixteenth to the nineteenth centuries. The idea that the book of nature as written by God complements the Bible as written by inspired prophets held hidden dangers for theologians, however. As science emerged to provide powerful observations about the workings of nature, people observed differences between God's two books. Some argued that differences between the Bible (God as written) and nature (God revealed through science) should be decided in favor of science, which was less speculative. Catholic authorities banned these publications because the arguments challenged the role of Catholicism's hierarchical interpretation of Holy Scripture. The work by Copernicus, Galileo, Newton, Darwin, and others presented serious scientific challenges to religious understandings of the world.

Science gained increasing independence from religious authorities, especially in Protestant countries. Science became less a vehicle to understand creation and more a vehicle to support state power and create wealth. By the twentieth century, science almost exclusively served the interests of economic development and a secular state. There still exist pockets of natural theology such as creation science and intelligent design, but most scientific study is no longer conducted for the purpose of finding evidence of God and insights into his plan.

RE-CREATE THE GARDEN

Sir Francis Bacon (1561–1626), the important Enlightenment author and architect of modern science, argued that science provides a way to finish God's work. Science and industry are divine gifts that should be used to create a man-made paradise. Assertiveness should replace humility. Thus, environmental scientists and managers intent on transforming nature are good shepherds doing God's work on Earth. Wild nature is inefficient, full of briars, predations, drought, and chaos, out of which human labor and ingenuity can create order and efficiency, finish God's work, and create heaven on Earth. Salvation can be reached through the hard work of domesticating nature and re-creating the Garden of Eden from which Adam and Eve were expelled.[23]

Gifford Pinchot (1865–1946), a pillar of the American environmental movement, preached the Bacon sermon. He was devoutly religious, even evangelical. At college he led religious services for his classmates and for many years thereafter avidly promoted the YMCA, which at that time was an evangelistic Christian movement. Pinchot preached progressive conservation with the conviction of a religious zealot. He believed God's kingdom on Earth would be realized in America through wise but aggressive use of natural resources, and he directed the tools of science and the power of government to secure a successful Christian nation and to create a new Eden on Earth.[24]

Among the first duties of every man is to help in bringing the Kingdom of God on earth. The greatest human power for good, the most efficient earthly tool for the future uplifting of the nations, is without question the United States; and the presence or absence of a vital public spirit in the young men of the United States will determine the quality of that great tool and the work it can do. This is the final object of the best citizenship. Public spirit is the means by which every man can help toward this great end. Public spirit is patriotism in action; it is the application of Christianity to the commonwealth, to the brotherhood of man, and to the future.

Pinchot was America's first native-born professional forester, the first chief of the U.S. Forest Service, a governor of Pennsylvania, a presidential candidate, and a principal architect of the nation's natural resource conservation policy. He saw professional natural resource management as a rational means to meet the resource needs of a growing Christian nation. The conservation ethic he championed brought science to the service of the Protestant ethic: avoiding waste and fixing inefficiencies. He implemented this philosophy as key adviser to President Teddy Roosevelt (1901–9). Together they changed the shape of the U.S. landscape, putting millions of acres of forested land under the charge of Pinchot-trained managers. These lands were scientifically managed for the greatest good for the greatest number for the longest time, which translates into economic development that could build and sustain "God's Kingdom" on Earth. Under the gospel of science, nature was managed to build the "City on the Hill." Pinchot had given natural resource professionals their calling.

WORK IS DIVINE

Perhaps the religious doctrine most influential over environmental conditions is not explicitly about nature, but rather about work and industry. Human ingenuity combined with a Protestant work ethic produced great change in the American landscape. Again inspired by John Calvin, Protestants generally, and Puritans especially, saw work and industry as divine. Idleness was a sin and work was blessed, so resource exploitation became a religious duty. The Puritans showed little appreciation for nature not turned to human benefit. They disapproved of recreation and aesthetic appreciation of nature; even hunting for pleasure was discouraged. Time and effort were best devoted to the worship of God or to the economic development of God's creation. In the words of Cotton Mather, a famous seventeenth-century Puritan, "What is not useful is vicious."[25]

Accumulation of wealth is an antisocial and immoral behavior in many cultures. These so-called gifting cultures grant status to those who distribute wealth, their prophets promise salvation to the meek, and those with the least are supposed to find the most in the afterlife. Protestant theology, in contrast, teaches that having wealth and possessions is not a sin for which one should feel guilty; it might even provide evidence of being predestined for salvation. Calvinists anxious about their destiny in eternity looked to signs of success in the here and now as indications of predestination. Max Weber, the influential twentieth-century social critic, argued that Calvin made accumulation a virtue out of what was previously a vice. Poverty no longer ensured God's good graces; instead, salvation required labor, diligence, and thrift. Protestantism converted prosperity from a foe to an ally of religion and in so doing aligned religious faith with capitalist zeal to create a force that changed Earth forever.[26]

God's charge to multiply and subdue the land, Protestant capitalism, and a dualist disrespect for nature combined to encourage the taming and development of all wild places. These forces coupled with technological advances fueled the Industrial Revolution, the demand for resources, and the materialistic culture that some environmentalists suggest is the root cause of current environmental problems.

APOCALYPSE

Hopes and fears of apocalyptic ecological collapse motivate a vocal and passionate minority at the fringes of religious and environmental communities. The end-times are momentous: Christ returns, chosen Christians find Rapture, those left behind suffer and die, creation as we know it is destroyed, and heaven and Earth are re-created free of sin and evil. Chapters in the Bible such as Daniel and Revelation are interpreted for clues about current events indicative of an increasingly imminent apocalypse. Websites monitor and interpret these events. The Rapture Index, for example, provides a "prophetic speedometer of end-time activity," and the Armageddon Clock "moves ever forward. Each tick of its second hand . . . brings the world nearer the starting point of what will be man's most horrific war." Indicators of how close we are to Rapture include the popularity of non-Christian religions, a faltering U.S. economy, moral "failures" such as gay marriage, the actions of Israel and Russia, and several environmental events such as earthquakes, drought, and global warming.[27]

Opinion polls show that 36 percent of U.S. residents believe the book of Revelation is true prophesy that predicts the end of the world as it will happen, 55 percent believe that before the world comes to an end the religiously faithful will be saved and taken up to heaven (i.e., Rapture), and 17 percent believe the world will end in their lifetimes. The popularity of apocalyptic visions is evidenced by *The Late Great Planet Earth* and other books by Hal Lindsey that have sold millions of copies worldwide. Also popular are books in the Left Behind series by Tim LaHaye and Jerry Jenkins, which often top bestseller lists. The Left Behind website presents images, testimony, and analysis that evoke a sense of urgency about becoming Christian so as not to be left behind in the imminent apocalypse. Membership in the Left Behind Prophecy Club offers "to help you understand how current events may actually relate to End Times prophecy."[28]

Environmental destruction plays a role in causing and signaling the Apocalypse. The book *Reckoning with Apocalypse: Terminal Politics and Christian Hope* describes how end-time reasoning has and should motivate conservative Christian environmental positions. It warns that ecological destruction has been hidden by Satan and is only now becoming obvious. The day of reckoning must be near because the extent of destruction is now becoming obvious:

> The masses of exploited, the hungry, the terrorized are beyond the view of most people in the rich countries. But as populations in the poor countries continue to increase dramatically while food production lags ever further behind, unrest and upheaval become harder to ignore. The build up of carbon dioxide for the greenhouse effect, the depletion of the ozone layer, and countless pollutants blighting life cannot be seen. Radioactivity from nuclear power plants and from other nuclear sites remains invisible. The killing of plankton in the oceans, which could eventually result in too little oxygen for humans to breath, is out of sight. Yet more and more the consequences of degrading the biosphere present themselves. Weapons of mass destruction are kept hidden. But each of them constitutes a plummet into the abyss of human evil. And within minutes they could bring vast disclosure of that evil. . . . Looming towards us is utterly desolating apocalypse.[29]

There are various subtle and not-so-subtle implications of apocalyptic beliefs for understanding nature and setting environmental policy. One implication is that environmentalists' concerns about pollution, climate change, and resource depletion can be discounted or dismissed because end-timers are confident that God will provide for human well-being until the end, at which point the chosen will be saved and Earth will be destroyed and/or renewed. If the end is imminent, then there is little reason why society should devote its limited energies to conserving natural resources, creating a solar economy, or restoring functioning ecological systems. Likewise, it is a waste of time to learn about recycling, wilderness, or sustainable agriculture. End-timers are fervent dualists. They focus on preparing their souls for Rapture and making sure that they and their loved ones will be among those selected.[30]

While Lindsey's books and the Left Behind series don't motivate conservation behavior, they may motivate environmental awareness, by pointing out environmental conditions that they believe provide important signals of Christ's return. Oil spills, global warming, and vanishing rain forests are suggestive of biblical apocalypse. As the seven seals are broken in the book of Revelation, the sun scorches humankind with fire and consumes all vegetation, earthquakes topple mountains, sea creatures die, and water is poisoned.

Another possible implication of the popular apocalyptic writing is a raised awareness about human action causing the Apocalypse. Some authors see God as the only possible cause of end-time, but other authors clearly see humans as a contributing factor. Pollution, technology, and selfishness stoke the fires of ecological crisis. God might perform miracles of destruction, but he also sets in motion human actions that destroy creation. Lindsey's *There's a New World Coming*, for example, argues that humanity brings catastrophe upon itself: "God simply steps back and removes His restraining influence from man, allowing him to do what comes naturally out of his sinful nature." Some apocalyptic extremists may even argue that we should accelerate environmental destruction in order to hasten the Second Coming and the glory it will bring to the chosen.[31]

Another implication of apocalyptic eschatology simmers within the secular environmental community. The science and alarmism of environmentalists fuel a secular apocalypse. Warnings about population bombs, mass extinctions, and global warming dominate environmental rhetoric and public understanding of environmental issues. Hollywood blockbuster films such as *Jurassic Park* (1993), *Outbreak* (1995), *Armageddon* (1998), and *The Day After* (2004) denote a sustained public interest in Earth-changing disasters and their causes, some of which are "natural" (i.e., asteroids) and some of which are self-inflicted (i.e., genetic engineering and global climate change).

Radical environmentalism such as that exposed by Earth First! has end-time threads woven into its stories. Its literature and philosophy are often pessimistic about an inevitable ecological apocalypse. Radical environmentalists fear that the industrial system is so entrenched and out of control that it cannot be changed by normal political means, so monkey-wrenchers resort to sabotaging visible and vulnerable parts of the capitalist, military, and industrial state system. Radical environmentalists' apocalyptic story also holds out hope and redemption: some life-forms, they hope, will continue after the environmental apocalypse and, if humans are among them, survivors will have learned their lesson and practice only primitive, bioregional, sustainable behaviors as they reconstruct society.[32]

Other Christian interpretations of the Second Coming reject any calls to neglect or degrade the environment; in fact, they find in it a renewed call for environmental stewardship responsibilities. Famed Reformation theologian Martin Luther (1483–1546), for instance, supposedly said: "If I knew the world was going to end tomorrow I would plant a tree." He was not a millennialist awaiting the Apocalypse. His faith and interpretation of Scripture told him to remain rooted in the belief that God so loves the world he would not destroy it and transport the church to some other place. Instead, God would come to live on Earth, with Jesus, in a time of peace and plenty. These interpretations of the Christian future advocate Jesus' teachings of supporting others, living simply, worshiping actively, and improving the lives of others around us and descendants who follow, rather than emphasizing death, destruction, and the end of time.[33]

CONCLUSION

Whatever the root cause, Christianity's complicity in conquering nature is hard to ignore. We can debate the Judeo-Christian tendencies toward dominionism, dualism, capitalism, and apocalypse, but must recognize the role that religion has played in settling, taming, and in some cases extinguishing many natures and cultures. Evangelists and missionaries walked beside conquerors and merchants, wittingly or unwittingly assisting in and justifying the replacement and subjugation of native landscapes. Likewise, religious teachings that encourage population growth and material consumption must accept that these behaviors have direct, profound, and lasting environmental impacts. In 1991 Nobel laureates and other respected environmental scientists widely circulated a critical letter addressed to the "American Religious Community." In it they complained about the apathy, if not complicity, of religious institutions toward contemporary environmental problems.

It is also hard to ignore that religious motivations have played a key role in many conservation efforts. People motivated by deep faith, such as Gifford Pinchot and John Muir, established in the United States one of the modern world's first and greatest environmental conservation and preservation efforts. Pinchot founded U.S. professional forestry and advocated progressive conservation in order to build and protect a strong Christian nation, a new Eden on Earth. He was motivated by religious faith and Protestant distaste for waste. John Muir, also drawing on Calvinist theology, emphasized God in nature. He strove to protect God's natural cathedrals and to baptize sinful, weary, urbanized people in nature's healing graces.

Today's God's Greens are mounting sophisticated theological rationales and practical actions to care for creation. Organizations such as the U.S. Catholic Conference, the North American Coalition for Christianity and Ecology, the Coalition on the Environment and Jewish Life, and the Evangelical Environmental Network advocate environmental awareness within mainstream religions. Most religious institutions now have formal policy positions on a wide variety of environmental issues, including energy conservation, animal welfare, sustainable agriculture, acid rain, oil exploration, and reducing materialism and consumption.[34] Justifications for these policies vary and include arguments to protect God's creation and to reduce human suffering, as well as reinterpretations of biblical texts that deemphasize dominion, reemphasize caring, and overcome dualism.

Religious-based conservation arguments also appear in mainstream environmental media. For example, the Nature Conservancy published "Viewing Nature as God's Creation: Common Interests between Faith Communities and the Nature Conservancy." *Outside* magazine profiled "God's

Greens," in which it describes environmental activists "armed with scripture and a righteous respect for nature . . . [who] have taken up the environmental fight and are waging holy war on behalf of an embattled creation." The secular National Association of Conservation Districts annually prepares and distributes millions of church bulletins and litany readings for religious services to support Soil and Water Stewardship Week themes such as "the gift of trees," "food for the future," and "in children's hands." The Sierra Club and the National Council of Churches recently teamed up to protest drilling for oil in Alaska's Arctic National Wildlife Refuge. They funded television commercials featuring charismatic nature scenes and a narrator talking about protecting God's creation.

These examples are just tips of large icebergs of faith. Enough international activity and support were present by 2000 to justify a United Nations review of worldwide religious programs promoting environmental agendas, which was summarized in an eighty-page publication. In 2002 the high-profile Worldwatch Institute published a similar review and argued for the potential of greater collaboration between religious and environmental institutions.[35] Environmental activists and philosophers are recognizing that religion provides people with a language and a rationale to think about nature *and* humanity. Few other cultural institutions are so ideally suited to deal directly with the deep moral questions about how to treat nature. Religion provides the ideals and principles that shape and guide many lives, offering a vision of the good life that is not based on consumerism, materialism, and self-interest. Religion strives to answer the big questions about the meaning of life and how we should treat others.[36]

Although the Bible says relatively little about humanity's relationship with nature, some of what is said comes very early in the text. In fact, God's very first words after speaking the Creation into existence are instructions about environmental ethics. God declared the Creation "good." Given the prominence where these instructions are placed in the Bible and today's constant talk of an environmental crisis, it is surprising that we don't hear more from the pulpit or the evangelical airwaves about environmental stewardship. Few social-political movements get far in America without an implicit if not explicit justification in Judeo-Christian teachings. Environmentalism is one of the few social-political movements to enjoy widespread public support for over a century. There seem enormous opportunities for religious and environmental faithful to join forces and further their common interests.

* 10 *
Human Nature

Humans Are Animals
More than Primates • *Nature versus Nurture*

Are humans natural? Or are we special, different from, and better than nature? These questions have befuddled us for millennia. Attempts to answer them can make the most articulate and educated among us appear inarticulate if not mystical. Ask people if they are natural and watch them struggle to answer. A brief flicker of uncertainty may cross their faces as two paradigms collide: (1) humans are different from nature as evidenced by religion, education, and space exploration; and (2) humans are connected to nature as evidenced by evolution, ecology, and anthropology.

Humans are, it seems, both part of and apart from nature. It is easy for someone growing up in modern culture to see clear distinctions between humans and nature. Language, government, commerce, education, art, and technology seem to set us far apart from our evolutionary cousins. The unpredictable, unforgiving, cold, and rainy natural world has been tamed with air-conditioning, roads, plumbing, and shopping malls. We insult people by calling them animals, implying that they are somehow less than human: ruthless, selfish, aggressive, and uncivilized. Yet as we learn more about the natural world, we realize our deep ecological, evolutionary, and emotional connections to it. And the more our technology and urbanization distance us from nature, the more we seem to yearn for its embrace. We are awash in cultural currents that simultaneously push us away from and pull us toward nature.

This chapter assembles evidence for both sides of the debate about human nature: (1) humans are part of nature and (2) humans are different than nature. The final section of the chapter explores the nature versus nurture arguments that often lie below the surface in debates over environmental policy. Some policies reflect assumptions that human nature is a blank slate that can be filled and shaped by culture, while other policies seem to assume that human nature is somewhat fixed, preprogrammed by evolution. Of course, different policies result from different assumptions. Deciding where humanness stops and naturalness begins is a major issue in assigning rights and inherent values to nature—the focus of the next chapter. Humans have rights—such as life, liberty, and the pursuit of happiness—that cannot be violated even for economic gain. Debate swirls around criteria used to distinguish humans, who have these rights, from other units of nature that don't.

HUMANS AS PART OF NATURE

The view that humans exist at the center of a universe created exclusively for them has been repeatedly challenged by scientific findings. Astronomy revealed that the sun does not revolve around Earth, that Earth is not the center of the universe, that billions of stars many times the size of our sun are so distant as to be only faint points of light in our night sky, and that these stars might support life on an infinite number of other worlds. Natural scientists such as Georges-Louis Buffon, Charles Lyle, and Charles Darwin became convinced by their empirical studies that the world was much older than suggested by Genesis. By the 1820s many geologists were certain that Earth's prehistory was a matter of millions not thousands of years. Paleobiology showed and evolution explained how countless species had come into existence, lived, and been obliterated, long before humanity appeared. Rather than viewing humanity as the purpose and center of creation, philosophers began to think about it as nothing more than an accident, inconsequentially located at the fringe of a grand universe. Humanity's fall from glory produced a bruised ego from which many of us have not yet recovered.[1]

The more we learn about biology, the more we realize humanity's deep connections to the rest of life on Earth. Comparisons among contemporary species show that humans and other life on Earth share the same genetic mechanisms of reproduction (i.e., DNA, RNA), the same basic cellular structures and cell functions (i.e., nucleus, mitochondria), and the same molecular biology that governs cellular housekeeping (i.e., ATP). The fossil record documents the gradual anatomical changes that evolved and connect all life back to 4 billion-year-old bacteria. For example, we share with fruit flies, and countless other past and present species, a gene (PAX6) that plays a key role in forming eyes. We share with other life-forms the same basic body shape (head, neck, body, appendages, symmetry, etc.) that emerged some 500 million years ago. And we share with chimpanzees, by some estimates, over 95 percent of the exact same DNA. Biology teaches us that we are natural: humans are a product of the same forces that shaped and operate all life on Earth.

We learn from ecology that everything is connected to everything else. The so-called web of life connects predators and decomposers, the aerobic and the anaerobic, air and water, plants and animals. Dramatic changes in one strand of the ecological web are likely to influence life on other strands, which in turn will influence still others. Human life, it seems, is dependent upon infinitely complex systems and interconnections that produce and maintain oxygen, soil, nutrients, and other life-support systems. Humanity now so dominates Earth that the structure and functioning of Earth's ecosystems cannot be understood without accounting for us. Our impacts on ecological systems are so extensive that they alter global climate, the distribution of life, and the function of life processes on Earth. Many branches of ecology now treat humans as just another factor affecting ecosystem processes.[2]

The Human Animal

The more we learn from our rational sciences about the workings of nature, the more the line between nature and humanity blurs and disappears. But perhaps what differentiates humans from nature is the very language, consciousness, passion, and rationality enabling this science? Alas, readers hoping for clear distinctions here will be disappointed. While we humans are exceptionally clever, we are not unique in this regard. Our cleverness differs from other animals only by matters of degree. Here I will discuss three aspects of cleverness that we likely share with other animals: awareness, feelings, and consciousness.[3]

Many if not all forms of life are aware of their environment. Much of evolution can be explained as the incremental refinement of awareness. Creatures able to differentiate among predators, toxins, nutrition, and shelter are more likely to survive and reproduce: plants seek light, amoebas move toward sugar, amphibians seek water and warmth, and herbivores avoid toxic plants. Sexual reproduction necessitates ever-greater levels of awareness. Members of the opposite sex must find members of the same species at the most likely time for conception. Elaborate displays of pheromones, colors, songs, and dance are used to advertise and assess potential mates.

Human awareness is neither unique nor superior—hawks see farther, canine olfaction is more sensitive, bats have remarkably more sophisticated hearing, and even crude machines such as thermostats are able to monitor and respond to minute environmental change. If humans are not unique in possessing awareness of the environment, aren't we special in our sensitivity to pain, pleasure, and other feelings these sensations produce? It turns out to be very difficult, and perhaps impossible, to objectively assess the quality and intensity of a feeling. Emotions occur within the minds of individuals, and thus are difficult to share and validate. Humans have the benefit of a sophisticated language and elaborate facial expressions, but still our inner experiences cannot be shared with others for objective evaluation. For example, headaches are events most people experience at some point in their lives. We know one when we have it. We don't like the pain, we try to avoid the suffering, and we can share a verbal description of it. We are able to observe behaviors indicating that others feel a similar pain, and we even can empathize with them. Yet we can neither directly feel nor verify another person's headache. Accessing and assessing the feelings of another species is even more difficult.

Do other creatures feel pain and pleasure the way we do? It is easy for those of us with pets to recall anecdotal events where our animal companions experienced intense feelings of pleasure, pain, or fear. Accidentally stepping on a pet's tail typically evokes squeals that we humans have no problem associating with feelings of pain. Threatening punishment for unwanted behavior not only stops the behavior but evokes cowering and retreat, suggesting fear and anticipation of pain. But anecdotal observations are an insufficient basis for answering fundamental questions about the very essence of humanity.

Let's look more specifically at pain. An inclusive definition of pain might include any event that an organism responds to by acting as if it is in distress. Most forms of life exhibit reactions that satisfy this definition. Trees, for example, have special biochemical responses to physical injuries caused by fire, insects, or mechanical abrasions. They excrete chemicals unique to the injury. Neighboring trees become aware of the chemical and respond in ways that prevent the same injury (i.e., they close their stomata through which chemicals and heat can pass into the tree and cause injury). In this way, trees not only respond to injury but communicate that they've been injured. No doubt these injury responses are automatic, inherited reactions that occur without thinking, much like we pull our hand away from a burning hot stove. However, responding to an injury does not mean that pain and suffering were *felt*. Feeling, at least in any sense humans can identify with, requires a nervous system that registers and responds to an injury.

Do organisms with nervous systems similar to humans experience pain the same way we do? Most animals respond quickly to prods of heat or electricity, often with vocalizations, physical contortions, defensive postures, and other behaviors that are similar to humans experiencing pain. Further, it turns out that most vertebrates share with humans similar chemical and neurological mechanisms that transmit and control pain. When poked, burned, shocked, or cut, they perspire, increase blood pressure, and release stress hormones such as adrenaline and steroids. Also in response to pain, their glands secrete increased quantities of endorphins and other natural painkilling opiates similar to chemicals released when humans feel pain. Reactions by mammals to painful stimuli can be suppressed using the same medications that suppress pain in humans (i.e., those used in dentist offices and outpatient surgery). Even the psychiatric medications used to treat human depression seem to have the same effects on dogs, cats, pigs, horses, and other animals, as is evidenced by their frequent prescription by veterinarians.[4]

Thus, even though we cannot directly observe the pain in other creatures, we can conclude with some confidence that humans share with some of them our reactions to painful stimuli. It still can be argued that we interpret these feelings in unique ways; for example, in addition to feeling pain, conscious humans may also experience grief, sorrow, and dread because of the implications and deeper meanings associated with the painful event. We don't just feel the pain of an injury, but we worry about the lost opportunities—a broken leg means missed hiking trips and soccer games as well as weeks of inactivity and inconvenience. If we argue that our interpretations of pain distinguishes us, then we must tread on the slippery and uncertain terrain that defines the ability to interpret situations—consciousness. Four qualities of consciousness are addressed here, all of which some animals seem to possess and use.[5] These include (1) the ability to respond flexibly and learn, (2) language use, (3) self-awareness, and (4) awareness that others are aware.

Nonhuman animal behavior has been shown to be complex and contingent, rather than simple, instinctive, or reflective. Many animals learn lessons and adapt to new situations rather than respond inflexibly, as would be expected if learning did not occur. Lab rats, for example, learn from others which foods have been poisoned. If experimenters are ethically inclined to do so, they can feed a rat poisoned food. Other rats quickly learn from observation which particular food made their fellow rat sick and thereafter avoid that food; that is, rats learn through observation. More elaborate examples of animal learning abound. They include everything from apes learning to wash sand off sweet potatoes or to extract termites from mounds with tools to teaching dogs to shake hands and roll over. But is the cognitive ability evidenced when a dog learns new tricks sufficient to support the cognitive flexibility required of consciousness?

Use of language provides strong evidence of consciousness. Even the most ardent skeptics of animal consciousness would have reason to pause if studies demonstrated that animals represent, manipulate, and communicate complex and abstract ideas. Not surprisingly, it turns out to be exceptionally difficult to prove that animals comprehend meanings associated with sounds or symbols. Sounds of alarm provide an example. Many prey animals squeal when they sight a predator, and nearby animals respond immediately with defensive behaviors such as hiding or fleeing. It appears as if the initial cry was actually a "warning" and that the others understood that "danger" was present. However, we may be imposing our human-based interpretation of the situation onto these actions. The same defensive behaviors could occur without the sound having any such *meaning*. The animals may be conditioned by learning or by evolution to evoke defensive behavior when the alarm sounds: those that flee live to pass on their genes. That is, the sound can evoke a response without conveying the meaning of "warning" or "danger."[6]

Vervet monkeys, about the size of small dogs, use different alarm sounds for different predators. Because each sound evokes a different behavior, it is tempting to conclude that the sound communicates meanings (more than automatically triggering defensive behavior). The warning sound for bird of prey provokes everyone to dive for cover under the nearest brush where birds cannot fly, while the warning sound for leopard sends everyone up trees and to the edges of branches where leopards cannot reach. The warning sound for snake, rather then sending monkeys scurrying into possible danger, gets everyone standing on their hind legs, looking at the ground for danger. Associating different sounds with different predators seems to suggest that the listeners understand the meaning associated with the alarm sound. However, it is still not obvious that the alarm sound was an intentional effort to warn other monkeys, which it would need to be if it were *meant* as communication. It is not obvious that vervet monkeys sounding the alarm observe and correct the behaviors of monkeys who misinterpret the signal. If they did, it would be easier to impute intentionality to the signal.

A prairie dog makes a high-pitched bark immediately upon seeing a hawk or an eagle. Is this bark an intentional warning, reflecting concern about the well-being of other prairie dogs, or is it an instinctual reflex that increases chances of survival? Nearby prairie dogs certainly respond to the bark as if they are in danger; they scurry for cover in burrows and brush. There is little doubt that the bark stimulates defensive behavior in other prairie dogs, but in order for the bark to be considered communication, doesn't it have to be made with some intent of sharing information? It is tempting for us to assume that the prairie dog intended to warn others of danger, but it seems more likely that the prairie dog had no such intention. Rather, the bark is instinctual. Prairie dog ancestors that made a high-pitched bark upon seeing a hawk were more likely to survive and pass on the genes that evoke the bark. The high-pitched bark not only disorients the hawk, but it also mobilizes many more targets for the hawk to choose from, as other prairie dogs become visible when they scurry across the field. The bark does alert others to the hawk, but it may have no *intention* of doing so. Rather than an intentional act of communication, it is a preprogrammed response that increases chances of survival and thus gets passed on from generation to generation.[7]

Better proof of language abilities in nonhumans may be found in the ability of some animals to assign meanings and to manipulate abstract symbols. Evidence of this sort comes from laboratory studies of primates. Chimpanzees possess most of the same cognitive abilities as humans. They seem capable of highly developed visual-manual skills, complex problem solving, tool use, toolmaking, and, discussed here, the ability to manipulate and remember abstract concepts. For example, human handlers taught a chimp to signal "more" to request more tickling. The chimp seemingly exhibits abstract thought by being able to apply that meaning in different contexts to ask for more hair brushing, more games, more food, and more entertainment.[8]

Further evidence of the ability to use abstract ideas comes from numerous match-to-sample experiments. Chimpanzees shown an item (say, a red ball) can be trained to match that item with an identical item located in a group of other items (say, one red ball and two red cubes). Some chimpanzees learn to match abstract qualities such as the color, size, and shape of objects. For example, when shown a red ball, they can match it with another red object, even if that object is a different shape or size (i.e., they select the red cube rather than the blue triangle or green star). It may take some time to teach them to match the first abstract quality, but once they've mastered it, they then learn much more quickly to match other abstract qualities (e.g., once they learn to match objects based on color, they are faster at learning to match objects based on size and shape). Thus, chimps seem able to generalize the lesson, which is a much more sophisticated type of learning. Even more impressively, chimpanzees are capable of making analogies. They can master the visual equivalent of this verbal analogy: small square is to large square as small circle is to large circle. They also seem to understand abstract analogies, such as a can opener is to a can what a key is to a padlock (i.e., an opener).[9]

Another criterion of consciousness is self-awareness. To be conscious implies being aware of yourself and of yourself in your environment—you realize that you are thinking about yourself, that you are both the thinker and the thought about. Some chimpanzees (and perhaps a few gorillas, elephants, and dolphins) seem to pass a common test for self-awareness known as the mirror-recognition test. When chimpanzees first encounter a mirror, they act as if their reflection is another animal, directing threatening, greeting, or other social responses toward it. After several days some act as though they realize the mirror is a reflection of them. They use the reflection to groom, observe, and manipulate visually inaccessible parts of their bodies. They pick teeth, make faces, blow bubbles, and manipulate food while watching the mirror. It seems as though they are aware and can think about their bodies.

In experiments to test self-awareness, chimpanzees are rendered unconscious by medication. Colors are then painted on their foreheads and ears, places visible only with the aid of a mirror. When the revived chimps encounter their reflections, they repeatedly touch the painted areas on their bodies, sniff and taste their fingers after touching the paint, and otherwise behave in ways suggesting that they understand it was their bodies that had been altered (by paint). Not all chimps demonstrate this awareness; some seem not to realize their body has been altered, and some never pay attention to the mirror in the first place. But those that did attend to the paint in their reflection arguably demonstrated some form of self-awareness.

Additional proof of consciousness comes from observations that animals are not only aware of themselves, but that they are aware that others are also aware. That is, some animals seem to think about what other animals think about. This ability is one of the most contentious in the human-nature literature; some researchers contend that the ability of humans to comprehend the intentions of others is the only quality that truly distinguishes humans from other animals. It enables the ability to anticipate, learn from, and teach one another. Creatures possessing this ability would have a mechanism to build culture.[10]

Deception provides a commonly cited example of this ability. To deceive someone, we must manipulate his or her awareness of our intentions—thus we must be aware that they are aware and that they, too, have intentions. For example, we could withhold or hide information so others do not know what we know or cannot observe what we are doing. Most examples of deception in nonhumans are anecdotal. In one case, a chimpanzee was observed walking by several grapefruit deliberately hidden by an experimenter. The chimp pretended not to notice the treasure, hence not attracting the attention of other, more dominant chimps. However, three hours later, when the dominant male and others were sleeping, this in-the-know chimp went directly to the hidden treasure and devoured it. Thus, the chimp seemed to be withholding information in order to keep the dominant chimps from eating the grapefruit.

Some of the more sensational (but still anecdotal) examples of deception involve sexual encounters among promiscuous chimpanzees. The dominant male aggressively discourages intercourse between his females and subordinate males. He brutally attacks any male or female that defies him.

Female chimps have been observed having intercourse with subordinate males behind boulders and in other locations out of sight of the dominant male. They also stifle their screams during climaxes with subordinate males, even though they still scream at climax during intercourse with the dominant male. It seems the females are withholding information so that they can gain sexual pleasures and avoid unwanted beatings. Subordinate males also practice deception by hiding their erections from the dominant male. An erect penis signals sexual interest and may evoke a beating. Its bright pink color contrasts sharply with the dark fur, making it an easy signal for the dominant male. Subordinate males with erections have been observed exposing themselves to receptive females but hiding the erection from view of the dominant male with an arm, body, or leg. These and similar anecdotal observations suggest that at least some chimpanzees possess an awareness of what others are thinking.[11] It seems that at least some animals can construct mental models of their worlds and their places in it. If so, then perhaps they can project beliefs, intentions, and hopes about their world and thus are similar to humans in this regard.

Despite ample anecdotal evidence, deception, like consciousness, is incredibly difficult to prove. It is instructive to consider an experiment that failed to find deception, even in animals given chances and rewards to be deceptive. Captive long-tailed macaques (monkeys) were given an opportunity to drink juice and avoid punishment if they practiced deception. They could obtain juice from two sources, one exposed to a human who was acting aggressively and squirting them with water whenever they approached the juice and the other location hidden from the human's view. The monkeys attempted to drink from the hidden source only about half the time, suggesting they could not fathom that withholding information by hiding themselves from the human would prevent the abuse.

Even though there is not a clear-cut case to be made for animal consciousness, evidence is accumulating to suggest that attributes once thought to distinguish humanity from the rest of nature turn out to be differences of degree rather than absolute differences. While human cognitive ability and self-awareness may far exceed what other animals possess, we share much with the rest of life on Earth.

HUMANS AS SEPARATE FROM NATURE

Most Western belief systems suppose that humans are special because we are physically, behaviorally, or spiritually unique. Supposed differences include the following: Humans are the only animals that fear death, possess a sense of humor, regularly practice deceit, anticipate one another's intentions, and actively teach one another. Humans are special because we think rationally, have science, and construct tools that can make other tools. Only humans are made in the image of God, have souls, and practice religion. Only humans cook food, eat it with utensils, and practice table manners. Only humans control fire and transform landscapes through its use or removal. Only humans can survive in all of Earth's environmental conditions and create cultural institutions to change ourselves. Perhaps most importantly, only humans negotiate an ideal and seek to create it in the real—that is, we plan our future.

These qualities supposedly don't just make us different from nature; they make us better. Almost by definition, qualities such as rationality, humor, language, free agency, and even polite eating habits are assumed better than irrationality, stoicism, ignorance, determinism, and vulgarity. These qualities have been used to justify the power and privilege of humans over nature. Throughout human history, similar arguments have been used to differentiate among humans: some people supposedly deserve fewer rights because they lack qualities that supposedly define human nature (i.e., non-white, less technology, non-Christian, homosexual, rejecting a modern worldview, etc.). Such differences have been used to justify war, colonialism, slavery, sexism, racism, and bigotry.

The logic applied to differences among humans is strikingly similar to the logic applied to differences between humans and nature. History has taught us that this logic when applied to human affairs has led to fallacious and self-serving justifications for the raw use of power. Thus we must be cautious and precise in using the logic of differences to justify our treatment of nature.

Judeo-Christian teachings distinguish humans from nature. Humans are different in that they are created in God's image. The divine soul links humans to God; it differentiates humans from the rest of nature more so than language, tools, or consciousness. The rest of creation, while important, is different from either God or humans. The Roman Catholic tradition links all of life on Earth to God in a great chain of being. The chain extends from God through the angels down to humans and on to animals, plants, and lesser aspects of creation. Life lower down on the great chain serves God through its services to life further up on the chain. The chain still distinguishes humans from nature but uses a hierarchy rather than a dichotomy: humans, being closer to God, are superior to other aspects of nature. Terminology such as "lower," "lesser," "higher," and "closer" clearly imply a moral pecking order. Protestant tradition is even more extreme than Catholicism in its promotion of separation by emphasizing an unbridgeable qualitative gap between humans and the rest of creation.[12]

The Enlightenment and modernism arose from the mystical and religious Middle Ages and also distinguished humans from nature. Bacon,

Locke, Galileo, and especially Descartes effectively severed the emotional umbilical cord connecting humans and nature. All nonhuman entities were, by definition, unconscious, unfeeling, and lacking free will. They were mindless machines whose behavior could be reduced to and explained by physical and chemical laws. Rationality was the supreme quality, the way in which humans resemble God. Not only does rationality distinguish humans from nature; it also enables and promotes the domination of nature so that human freedom and happiness become possible.

Humanism extends the Enlightenment and modern projects in a direction that further separates humans from nature. Human science, rationality, and ingenuity take center stage, replacing God as the navigational guide for charting social progress. Humanists believe that science and rationality provide self-corrective methods of critical thought and thus provide the best hope of resolving current problems and charting humanity's course into safe harbors. They question the capability of religions to offer moral guidance, pointing to wars and terrorism motivated by religious differences, as well as moral lapses of religious leaders. Logic, critical thinking, and experimentation, not religious doctrine, provide the means to develop moral principles and ethical values.

Humanism attempts to sweep away the superstition, mythology, nationalism, and fundamentalism that counter rationality. Religious and ethnic loyalties are seen as constraints. There is no place in humanism for a theistic God or unquestioned mystical traditions. Truth, albeit partial and evolving, is sought through rationality and science. Humans become stewards of their destinies on Earth, and their behavior is guided by rationality, not tradition, divinity, or genetics. Individuals think for themselves and wrestle control over their lives, while education instills integrity, trustworthiness, benevolence, and fairness in all people. "Good" societies are those that nurture liberty, equality, democracy, privacy, and self-determination. Nature therefore becomes a vehicle for humans to achieve their ends and matters only to the extent that it affects the human condition.[13]

The postmodern and constructionist paradigms extend humanist logic and further separate humans from nature. Human nature emerges unfinished and unconstrained by biology. Language, science, and culture—human symbols—ultimately define humanness, human goals, and human limits. A person's identity is shaped by the local and historically situated cultural context in which the person participates, not by evolution and genetic inheritance. Because culture is temporally and spatially specific, constructivism suggests that there exist no essential or shared qualities of human culture—or human nature. Rather, infinite possibilities emerge from combinations of environmental, cultural, social, and linguistic conditions that vary and change through time and space.[14]

The religious, humanist, and postmodern worldviews differ in their contentions that human nature is divinely inspired or culturally constructed. However, they are similar in their implications that humans are special and distinct from nature. Humans possess either God-given or constructed realities; nature is secondary and dependent upon human understandings.

Beyond Primates

Despite being genetically similar and evolutionarily related to all life on Earth, humans possess a relatively large brain, a more sophisticated speech apparatus, and other anatomical qualities that distinguish us from our closest evolutionary cousins. Moreover, the cognitive distance between apes and humans appears to be even greater than might be imagined from comparative anatomy. Language provides humans unparalleled ability to represent objects symbolically and to manipulate and share these symbols. Humans use symbols to manipulate and share their complex internal representations that otherwise have no external form in the natural world. The occasional chimp may be able to signal for "more" food and "more" tickles, but it remains controversial whether they comprehend the meanings that humans assign even very simple abstract concepts such as "more," or if they merely learned through trial, error, and rote memorization that the combination of "more" and "banana" produces additional bananas. By analogy, consider your rote memorization of the multiplication tables: for example, $4 \times 6 = 24$. The correct answer can be, and often is, learned and repeated without consciously thinking that four sets of six units produce twenty-four individual units. In other words, the meaning need not be understood to produce the "correct" answer. Thus, some researchers argue that even laboratory language accomplishments such as "more" tickling do not qualify as language.[15]

Humans grasp language with ease and interest. Children spontaneously name and describe objects they see: "tree," "car," "bird." They do this even without showing interest in the object and without seeking any reward. They also spontaneously describe these objects by using adjectives: "big" tree, "fast" car, "red" bird. Chimps, in contrast, must be painstakingly trained with careful reward structures to communicate using abstract symbols. They must be kept hungry so that they are motivated to pay attention during training sessions. Moreover, the symbols they use are not strung together into sentences to ask or answer abstract questions. While chimps can be taught in laboratory settings to use several dozen symbols, they do not use abstract symbols in the wild. And, as noted in the previous section, it is even debatable whether the warning sounds of animals in the wild, such as those by vervet monkeys and prairie dogs, count as language because there is little evidence of intention.

At a very early age, humans grasp that other humans have intentions and are self-aware. We manipulate one another's attention with gestures, words, and actions. Nonhuman primates do not. They do not point or gesture. They do not bring others to locations where interesting things can be observed. Nor do they actively offer objects for others to consider. Perhaps most importantly, they do not intentionally teach new behaviors. Other animals learn through imitation but do not teach. Imitation requires a novice to observe an expert and often produces a flawed copy. Teaching requires intention, reverses the flow of information, and ensures the lesson is learned by correcting any flaws. For example, a clever Japanese macaque snow monkey learned that dipping a sweet potato in water cleaned it of unwanted sand better than did brushing off the sand by hand. The behavior slowly expanded throughout the troop as others learned this trick by imitation. However, there is no evidence that monkeys in the know actively taught others how to clean potatoes or corrected incorrect potato-cleaning behaviors. Monkeys that ate sand-covered potatoes or dipped their sandy potatoes in the river mud rather than the deeper, clearer water were not corrected.

Stated differently, nonhuman animals appear not to intentionally manipulate the attention or purpose of their peers and thus do not actively teach one another. They learn only through direct involvement, experimentation, and, at best, imitation. Learning by watching how others solve problems requires understanding that others are trying to solve a problem, which only humans seem to do well. Teaching demands more. It requires comprehending that others want to learn. Only humans seem to possess this teaching ability that enables social learning and cultural evolution.[16]

Human language ability puts additional distance between us and other life on Earth.[17] Abstract language allows us to debate right and wrong, and thus evaluate behavior using more than the instinctual calculus of natural selection. Ideas such as rights, abuses, future, and mortality inspire different behaviors than does survival of the fittest. The ability to communicate abstract ideas creates opportunities to construct a moral code of conduct, with which comes responsibility. If, for example, we agree that freedom, democracy, equal opportunity, dignity, and pain-free existence are morally good things, then we have a responsibility to provide them by abolishing slavery, oppression, and tyranny. Perhaps this responsibility, more than any unique abilities, is what most distinguishes humanity from the rest of nature.

Are humans born empty and malleable, with blank memory files available for downloading cultural lessons and social norms? Or are our personalities, intelligence, and behaviors preprogrammed by evolution and controlled by genes? If the latter case, do these lessons predispose humans to be environmentally destructive or have we inherited a love and respect for the nature with which we evolved? These questions prove as intractable as questions about whether humans are natural. They have been debated for generations and remain topics of lively debate and study. The more we learn about human nature, the more it seems that both nature and nurture determine our potentials. The purpose of this section is not to decide between nature and nurture but rather to illustrate how assumptions made about these issues affect environmental policies and preferences.

"That's human nature!" is a common explanation of everyday behavior, implying that some human activities are excused or expected as instinctual responses to environmental stimuli. This fixed human nature supposedly makes us innately good or bad, competitive or cooperative, aggressive or peaceful, matriarchal or patriarchal. Theistic explanations suggest that God gave us these traits to help us find our way in this world. While theistic explanations are important, the focus of this chapter is on secular explanations of human nature. Evolutionary psychologists suggest that our traits and personalities reflect evolutionary lessons that increased chances of survival and thus are programmed into our genes. People exhibiting certain tendencies were more successful at reproducing themselves, and thus the genes enabling these traits were passed on to future generations. Inherited traits might include selfishness, territoriality, pair-bonding (marriage), empathy for family members, incest aversion, and inventiveness. Research on identical and familial twins raised in the same and different environments, for example, demonstrates that upward of 50 percent of intelligence and personality is inherited.[18]

Attempts to tease apart nature and nurture are difficult, at best, and studies often produce ambiguous findings.[19] Consider, for illustrative purposes, arguments about whether humans are individualistic or communal.[20] From the individualistic perspective, humans are seen as selfish individuals who seek resources, mates, and status—tendencies selected from our long history of hunting and gathering in the savanna. They help explain our species' propensity for competition, war, slavery, colonialism, and sexism. They are used to justify free-market competition (as discussed in the chapter on evolution) and the need for environmental regulation through coercion. If people are assumed naturally selfish and destructive, then we must restrict and restrain our natural tendencies, which will eventually destroy Earth.

An alternative interpretation of human nature suggests that humans are essentially cooperative and communal, tendencies selected from our long history of codependency for food, shelter, and care. Individuals that cared for the community, family, or tribe were more successful at reproducing and passing on their genes to future generations. These tendencies supposedly are thwarted by the individualistic and competitive qualities of contemporary culture. Thus, so goes the argument, we need to emphasize community over individual rights and a socialist model of governance instead of free-market capitalism. If people are assumed to be cooperative, then we might choose to implement community-based, cooperative strategies to resolve environmental problems. Presumably people would willingly work together to protect environmental quality because doing so would be good for the community.

Biophilia goes a step further and argues that our genetic inheritance motivates a love of nature that might encourage environmental protection.[21] It contends that humans possess affinities for nature because we evolved in nature. Humans are predisposed to like verdant, water-filled, easily navigated landscapes because these conditions facilitated survival. Landscapes that evoke negative reactions provide no sustenance, hide predators, and offer no easy means of escape. Humans who avoided these landscapes were more likely to survive and reproduce. Urban blight, congested highways, denuded forests, and polluted waters, therefore, evoke stressful and negative reactions in modern humans because we are predisposed by evolution to avoid them. Humans should be inclined to prevent or repair these problems.

Biophilia also argues that our language, logic, and sanity require contact with nature. At specific points during every person's maturation, one needs to observe natural cycles, apply natural metaphors, participate in natural events, and otherwise learn lessons from nature. Without these stimuli, we are less whole, less human, and less sane. We should therefore be motivated to preserve healthy natural settings and the opportunities to experience them.[22] Thus, biophilia suggests that our genetic inheritance should motivate considerable sympathy and empathy for nature and that we should preserve natural spaces so that we remain healthy and sane.

Seemingly in conflict with beliefs that genes influence our choices, yet coexisting with it in our culture, is belief in a blank slate.[23] Many people assume that they create their own opportunities through hard work and by selecting appropriate experiences such as a university education and the right friends. Rational deliberations of lessons learned from these experiences shape our destinies, not automatic responses preprogrammed millions of years ago. Education, advertising, parenting, and prisons assume that culture and learning shape human behavior, steer cultural evolution, and cure social ills.

Blank-slate philosophies of human behavior assume people are born absent of any serious direction from nature: culture fills blank minds with symbols, rationality manipulates these symbols with logic, and human culture dominates evolutionary inheritance. Biological evolution created a human mind that enabled cultural evolution, which now outpaces and outclasses the force that birthed it. The human mind is flexible and creative and allows humans to adapt to almost any condition on Earth and beyond. The future is not limited by the dead hand of the past; cultural evolution frees humanity from the limits of biological evolution. Cultural institutions fundamental to modernity—such as literacy, science, and democracy—assume some level of cultural determinacy is possible, if not preferable.

A blank slate would be receptive to any number of environmental ethics, wasteful and dominant or frugal and humble. Blank-slate philosophies argue that we use our rationality to identify and negotiate an appropriate land ethic and that we use cultural institutions to instill and enforce this ethic in human agents.

CONCLUSION

Human nature is difficult to discuss because humans are at once apart from and a part of nature. We are made of the same stuff, rely on many of the same processes to accomplish cellular housekeeping, survived related tests of natural selection, and may even think and feel similar thoughts and feelings. Yet we are also different. We can communicate with the future using both genes *and* culture. We can manipulate symbols in our large brains and share those symbols with those present now and those yet to come. Our self-awareness and creativity allow us to shape our own potential in ways not available to our evolutionary cousins. We have birthed a human culture with as much power to shape life on Earth as the processes of evolution that birthed us. We are both special and natural.

Our assumptions about human nature influence our environmental policies. Defining humans as separate from nature may invite, or at least make easier, the abuse of nature. The otherness of a thing provides reason for all kinds of abuse, from simple neglect to destruction. The presumed otherness of humans—Aborigines, blacks, Christians, homosexuals, Jews—has historically justified their oppression and abuse. Treating nature as other—as different, foreign, or strange—excuses us from extending it the rights, compassion, or concern we willingly extend to those of our own kind. Human specialness need not necessitate environmental destruction, however. Recognizing our specialness creates a sense of responsibility to develop and implement a moral code that considers and respects other life on Earth.

Recognizing our naturalness also might encourage respect for nature. Human conduct toward nature might be held to a standard similar to human conduct toward humans—do unto nature as you would have done unto you. Animals change from brutes and beasts to fellow creatures, companions, and even cousins. All of nature, including the soil, becomes part of a community in which humans are but one member.

Ultimately, none of this nuanced reasoning may matter. The naturalness of humans is too often thoughtlessly used as a rhetorical trump card. It excuses promiscuous, competitive, and greedy behavior as well as communal, caring, and nurturing behavior. It justifies craving contact with nature as well as the domination and destruction of nature. That is, it justifies just about every position on the ideological spectrum.

Few arguments are more frustrating and paralyzing than those in which naturalness is played as a trump card. One person argues that humans are unnatural and that nature is good, and therefore any human action necessarily degrades environmental quality. Another person argues that humans are natural, and if nature is good, then any human actions are as justified: human bulldozers and sky cranes are no different than the works of beavers, ants, and elephants. All creatures modify their environment. It is natural to do so. If the beavers do it, so should we. On and on the arguments go, trump card played atop trump card, spiraling toward paralysis. Most of these trump cards end negotiations in exasperation and hostility. Negotiators are unable to go beyond deep-seated, unspoken assumptions about human nature.

The point I'm making is that both protection and destruction of nature can be justified by assumptions about humanity being a part of or apart from nature. We've been debating these issues for centuries. The same arguments recycle, and the chief result is paralysis. Rather than waste our precious time trying to resolve whether humans are natural (and if they are or are not, what that means for environmental policy), let's find another way to think about the problem. Which natures do we want? It is time to choose. We can no longer afford to be deadlocked in rhetorical paralysis.

☀ 11 ☀
Rightful Nature

Source of Rights · What Gets Rights?
What Rights Does It Get?
Biocentricism versus Ecocentrism
Hunting · Vegetarianism

During a typical April when many of us hide Easter eggs and give each other marshmallows shaped like chickens, over 700 million live chickens will be slaughtered, their carcasses destined for frying pans and soup cans. That's nearly 25 million "broilers" or "fryers" killed each day of the month, 1 million each hour, 17,000 each minute, 287 each second; several thousand in the time it takes to eat a small chocolate egg.

Chickens now grow to the marketable size of five-plus pounds in eight short weeks. Neither the animal nor their roosts resemble the romantic images of small family farms with chickens milling about pecking for food. Generations of selective breeding for valued white meat have created awkward, lopsided chickens that have skeletal problems and trouble walking. Industrial warehouses can hold twenty-five thousand or more birds in conditions so crowded that the living stock will suffocate from the heat and stench if ventilation systems fail. Lighting and growth hormones are manipulated to maximize food intake and weight gain. Chickens are treated like machines for making meat with little or no concession to any aspect of their welfare except those that affect meat production.

Birds headed for "processing" are deprived of food and water for many hours because a "last meal" does not affect meat quality and therefore makes no economic sense. They are crammed into small cages and trucked great distances in all manner of weather, some dying from exposure. In the last few minutes of life, the bird is hung upside down from leg shackles on a long conveyor belt leading into the slaughterhouse. An electrified bath knocks the bird unconscious, loosens its feathers, stops it from flapping, eases processing, and reduces secretion of stress hormones that create undesirable properties in meat. Arteries in the neck are then cut (but not the nerves so that the heart keeps beating) and the chicken bleeds to death for about ninety seconds before being dipped into scalding water to further ease feather removal. Occasionally some birds enter the boiling water alive because of the difficulty in cutting necks.

Meat chickens may lead harsh lives, but they probably fare better than their egg-laying cousins. Nearly 80 billion eggs are laid for human consumption in the United States each year. More than 10 billion more are laid to become either layers themselves or the broilers just discussed. Today's hens are super layers, averaging over 250 eggs per year (compared to the several dozen by their wild ancestors).

Hens spend their lives with five or more other chickens in a small wire cage. Conditions can be so crowded that hens are unable to stretch their wings, preen themselves, or turn around. Smaller birds that fall on the wire mesh floor can be crushed to death and cannibalized. Urine and feces rain down from cages stacked in tall towers. To prevent chickens in such close quarters from pecking and clawing each other to death, it is routine to clip off their beaks and amputate their claws; both procedures apparently cause pain but anesthetics typically are not administered. Close quarters and, perhaps, neuroses cause some hens to rub their bodies raw against the wire cages. When egg productivity tapers off, the hens are deprived of food and water for several days to intentionally stimulate molting—one last bout of egg laying. A hen can live for ten to fifteen years, but under factory farm conditions they are expected to die before their second birthdays. Ten to 15 percent of layers are unable to cope with the conditions and die even sooner. The bodies of hens too exhausted to lay eggs are sold to slaughterhouses.[1]

Beef, veal, dairy, and pork production also emphasize economic efficiency rather than animal welfare. These practices, refined over centuries, deliver the amazing variety and abundance of inexpensive foods that support a growing human population. Medical science also compromises animal welfare. Animal lives have been sacrificed, for example, in studies to find cures for polio, smallpox, rabies, tetanus, and other dreaded human diseases. Pigs, dogs, and other large mammals are regularly used to train surgeons. Their vascular and digestive systems are similar enough to humans' to allow doctors to develop and test experimental procedures. A kidney transplant pioneer was once asked why he experiments on animals. He explained that a majority of his first subjects died. He revised the techniques and tried again. This time a majority of the patients survived. In the next trial, only one or two patients died. In his fourth group, because of lessons learned on previous subjects, all transplant recipients survived. His point in telling this story is that the first three groups were dogs; the fourth group consisted of human babies.[2] In the near future, animals will be more than just training devices and experimental subjects; they will grow replacement body parts for humans. Pigs, for example, are currently being genetically engineered to grow humanlike teeth, arteries, kidneys, and livers. If that technology succeeds, these parts will be harvested and the lives of pigs sacrificed in order save or improve human lives.

Cosmetics are sometimes tested on animals to ensure they are safe for human use. The Draize eye irritancy test is one of the more notorious procedures. It evaluates damage to eye tissue. Rabbits are bound so that they cannot paw their eyes, which may be held open with tape or clips. The deodorant, mascara, or other test substance is placed in their eyes, and reactions are observed for as long as several weeks. Rabbits do not have tear ducts to clean the irritants, so the reactions are likely to be more extreme than anything humans experience. Such tests are disputed and controversial. The leading U.S. trade association for personal care products has position papers recommending alternative test procedures, but some of these options are deemed too costly or insufficient, so many cosmetic, glamour, and hygiene products continue to be tested on animals.[3]

While you may not directly burn, brutalize, or boil animals alive, your lifestyle might support these actions. What happens behind the scenes in order to support our lifestyles has not gone unnoticed. Organizations such as the Animal Liberation Front (ALF) consider these behaviors so reprehensible that they justify direct actions against organizations and people deemed guilty of abuse. For example, persons claiming to act for ALF take credit for targeting banks, research labs, factory farms, fast-food restaurants, pet stores, and many other locations accused of abusing animals. Attacks involve, among other things, breaking windows, setting fires, and releasing animals. Some more radical movements target employees of organizations thought to use abusive animal tests. They vandalize workers' homes and even threaten physical harm. In one dramatic example, booby-trapped letters were mailed to animal researchers, hunting guides, and other targets. The note read: "Dear animal killing scum! Hope we slice your finger wide open and that you now die from the rat poison we smeared on the razor blade." Even ALF members debate whether violence and vandalism are effective and warranted, but few doubt the need to act. Opponents call them terrorists; proponents call them revolutionaries.

> ...the desire to make a fundamental change in society by throwing oneself against the atrocity of animal abuse...is the right thing to do. The debate over the ALF has never been a question of what is morally justified. How to best bring about change, though, is open to debate. Everyone involved in the animal liberation movement has doubts about the effectiveness of their actions and is searching for the best way to fight animal abuse and exploitation.... ALF

actions are dynamic and inspirational. The ALF can interrupt the dreariness of everyday campaigning with drama that reveals the animal rights struggle at its most essential level, if only for a short time. ALF actions can be a symbol of the revolutionary potential of our movement.[4]

Laws against cruelty to animals exist in all fifty of the United States. These laws do not give animals legal rights, and there is a wide range of definitions about what constitutes abuse, but the laws do provide some protection against deliberate human actions that cause excessive suffering in animals. Europe has gone further. In 2002 Germany amended its constitution so that "the state takes responsibility for protecting the natural foundations of life *and animals* in the interest of future generations." In 1992 Switzerland recognized animals as *beings* and not things. In 2003 the European Union banned the sale of new cosmetics developed using animal testing. Polls regularly show that a majority of Americans believe undue pain and suffering should not result from human uses of animals, but we are split on whether animals should be used just for food and medical research or also for clothing and cosmetics.[5] American industry is responding to these concerns. For example, multinational corporations such as McDonald's, Burger King, and other fast-food giants have forced producers of beef and eggs to implement more humane procedures for raising and killing livestock, and fish-processing companies now require that tuna be caught in ways that allow dolphins and turtles to safely swim away.

Most of us don't choose the radical path of action advocated by ALF, but we are nonetheless troubled that our lifestyles produce pain and suffering in animals. We cannot deny that our living inflicts pain and death on other creatures. Even vegetarians must eat plants that were once living, and all of us depend on the bacteria in our guts giving up their lives so that we may digest not just the sugars they excrete but their very bodies. Environmental ethicists debate these issues and suggest reasons and rationales that may help us develop a personal philosophy consistent with our behavior.

WHERE DO RIGHTS COME FROM?

It is common in environmental ethics to distinguish among instrumental values, inherent values, and rights.[6] Rights and inherent values exist *in* nature independently of humans, or at least independently of any use, experience, or appreciation that humans may have *for* nature. Instrumental values *for* nature, in contrast, are assigned to things affecting human welfare and happiness. Something has instrumental value if it is used as an instrument or means to produce food, shelter, health, profit, fun, or other qualities that humans desire.

General Grant, the third largest tree in the world introduced in chapter 1, overflows with instrumental value: humans value it for the houses it could build, for the heat it could produce if burned, as well as for the symbolic meanings and emotional reactions it evokes. Building material and heat are obvious instruments that serve human needs, but so are symbolism and awe-inspiring size, which satisfy some human desire for meaning and emotion. General Grant is probably of greater instrumental value alive and thriving than dead and converted into neat stacks of two-by-fours. People probably gain more benefits from the symbolic and aesthetic appreciation than from the economic benefits of building materials. Most trees suffer different fates because of this instrumental calculus; for most trees, the economic value of their fiber is greater than the instrumental value of the meanings and emotions they evoke. Arguably, General Grant also has rights. It is alive. It grows, seeks resources, and resists infections. By these actions of self-organization, the tree demonstrates that it has an *interest* in living.[7] Killing the tree could be interpreted as ignoring the tree's interest in living— committing the human equivalent to murder.

Respecting the rights of nonhuman life generates instrumental value if doing so makes us feel good about ourselves. Nature provides a mirror into which we look to learn about ourselves (see chapter 13). What do we learn when we look at the way we treat animals raised for food or used in experiments to test cosmetics? Mahatma Gandhi said in defense of his vegetarianism: "The greatness of a nation and its moral progress can be judged by the way its animals are treated." We can learn about ourselves, and feel good or bad about what we learn, by looking at how we respect the rights of animals and other aspects of the world around us.

Who among us deserve rights: Rich men with property? White men? All men? Women? Jews? Homosexuals? Children? Immigrants? Convicts? Fetuses? Stem cells? Primates? Mammals? General Grant? Those lucky enough to possess these rights are protected, to some extent, from exploitation. Rights cannot be reduced to an economic or instrumental calculus. If rights are to be violated, then the violator must pause and provide good reasons. The fate of things without rights lies solely in the harsh calculus of instrumental values. If a bunny has no rights, we are free to test our cosmetic products in its eyes as long as some humans value glamour more than they value the pain and suffering of bunnies. If General Grant has no rights, we are free to cut it down as long as the value of its wood exceeds the value of the aesthetic and moral experiences it evokes in humans.

It is assumed in the discussion that follows that the concept of rights is a human fabrication, a consequence of the way we think and act, a part of our worldview.[8] If humans did not exist, the concept of rights would not exist. Admitting that rights are assigned by humans need not lessen their importance or influence. Human rights, after all, have been negotiated, applied, and refined over time. The rights we grant ourselves are not fixed, essential properties. We continue to debate and change our ideas about which humans deserve rights (white males, women, ethnic minorities, etc.) and our ideas about what rights are deserved (happiness, privacy, property ownership, freedom of speech, democracy, education, health care, etc.).

Advocates of nature's rights argue that society is on a logical path of expanding our moral community. For a long time, only white, straight, pious men with property enjoyed all rights. Gradually modern cultures enlarged the moral community to include women, people of color, atheists, and homosexuals. To do otherwise would now be considered sexist, racist, and biased. The next logical extension of rights, so goes the argument, is to include primates, mammals, other animals, and perhaps ultimately plants and the larger ecological community.[9] Before we extend these rights and privileges that demand respect and protection, we need to ask two hard questions about where we draw the line on inclusion: (1) What units of nature deserve rights? And (2) what rights do they deserve?

WHAT UNITS OF NATURE DESERVE RIGHTS?

Various criteria have been proposed for separating those that deserve rights from those that don't, including skin color, language, intelligence, feeling pain, and willingness to live (see table 2, column 3). But as was discussed in chapter 10, humans are not unique in possessing these qualities; other units of nature differ from us only by matters of degree. It is hard to find a criterion we can apply with consistency. For example, if we grant rights to human infants and invalids that possess minimal cognitive and emotional abilities, why don't we also grant rights to chimpanzees and dolphins possessing greater amounts of these same abilities?

When answering this question, try to avoid speciesism, which is the environmental equivalent of racism and sexism. That is, avoid using arbitrary criteria (i.e., race, gender, and species) to hide power and political agenda. The color of human skin, for example, is and has been used to separate humans into classes of owners and property, civilized and barbaric, developed and developing. Likewise, gender is and has been used to limit economic and political potential. Most people now recognize that skin color and gender are irrelevant to decisions about many of the rights, privileges, and potentials of humans. Race and gender were just convenient and sometimes unwitting smoke screens that justified the status quo and kept power and opportunity under the control of white men.

Speciesism uses the same logic as racism and sexism, but applies it to bio-rights. It uses arbitrary criteria to defend the rights of humans to exploit animals and other units of nature. It employs the following circular logic: Humans are superior because humans are different. Or, stated differently, other creatures are inferior because they are not human. Speciesism supposes that distinguishing qualities of humanness (i.e., language, consciousness, culture, etc.) make us better and deserving of special rights and privileges. This argument is arbitrary, as becomes painfully evident when drawing a parallel to racism: it is similar to suggesting that white people enjoy rights not available to people of color because they are not white. Speciesism arbitrarily makes the qualities possessed by humans *the* criteria by which all other species are measured.

Table 2 (column 3) illustrates several qualities we might use to decide which units of nature deserve rights. Let's examine several table entries as a way to explore issues more fully. Animals such as primates, elephants, and dogs typically enjoy more protection than other forms of life. Why? Perhaps it is because people more easily empathize with their expressive faces, familiar body shapes, large size, and understandable behaviors. Collectively these qualities make an animal charismatic. Wolves, bears, and monkeys have it; beetles, reptiles, and bacteria don't. Charismatic creatures evoke in humans an aesthetic sense of awe and relatedness, and motivate numerous organizations to provide these animals with shelter, care, and protection. To argue that charismatic animals merit rights, however, is an example of speciesism.

Look near the bottom of column 3, at some of the more inclusive criteria listed. Most life-forms possess an interest in living. Arguably anything that acts in ways that prolong its survival and that avoids annihilation has an interest in living and, thus, perhaps a right to behave in that way. Elephants, insects, and fungi are born, mature, and reproduce—demonstrating a purposeful if not intentional interest in living. To hinder that interest or restrict that potential might be considered a violation of their rights. From this moral perspective, all living creatures have a right to their lives. Assuming we can agree on what it means to be alive, this criterion provides a clear place for us to draw the line differentiating those that have rights from those that do not. Yet it is not a pragmatic criterion. How can we use it to help make decisions? It is difficult to live without taking life; even our digestion requires the death of countless *E. coli* bacteria.

There exist even more inclusive criteria than the interest in living, and it is instructive to consider them. We could extend rights to anything with latent potentials. Tree roots, for example, demonstrate their potential by seeking moist, aerated, nutrient-rich soil. These potentials would be frustrated in the dry, compacted, oil-soaked soils found along urban streets. Does this make urban forestry an ethically questionable activity? Similarly a gene has the potential to produce life and reproduce. Unborn generations have untold

Table 2: What Deserves Rights and Why?

Which Units?	*Which Organisms?*	*Which Qualities?*	*Which Rights?*
Earth	humans	soul/sprit	life
Gaia or biota	primates	skin color	reproduction
ecosystems or	dolphins	charisma	healthy diet
communities	elephants	size	exercise
mountains and rivers	dogs	rationality	minimal pain
species	birds	language	happiness
populations	reptiles	self-awareness	standing in court
organisms	insects	pleasure seeking	health care
soil	trees	pain avoiding	free speech
organs (e.g., heart)	grass	autonomous vs.	vote
cells	weeds	instinctual	trial by peers
DNA	fungus	interest in living	practice religion
genes	bacteria	subject to evolution	bear arms
rocks and ice	viruses	latent potentials	

It is difficult to defend which units of nature (column 1) or which organisms (column 2) deserve rights without being arbitrary and guilty of speciesism. Column 3 lists a range of reasons or criteria that can be used. For example, all organisms possessing a certain quality (say the ability to use language or feel pain) could be granted certain rights.

potentials. Even nonliving things have potentials. A car requires gasoline to realize its potential. A toaster only reaches its potential when supplied with bread and electricity. Instant coffee lies dormant until mixed with hot water. Coal has the potential to become a diamond. Admittedly these arguments quickly become absurd, but they serve to illustrate the slippery slope on which we stand when seeking defensible criteria to justify which units of nature deserve rights.

Even the capacity to experience pleasure or pain—or, more subjectively, happiness and suffering—fails to provide a clear and foolproof criterion for granting rights and protections. It is tempting to argue that anything that feels pain should have the right to minimize pain. Many organisms avoid pain and seek pleasure, and we all know from firsthand experience that pain is bad. A consistent policy that tries to minimize pain proves problematic, however, because it is impossible to live life without causing/experiencing pain. Learning and growing cause pain or discomfort. For example, the short-term discomfort caused by discipline and delayed gratification motivates some of humanity's finest achievements. Fine art, scientific discovery, and athletic prowess are wonderful accomplishments that require enormous sacrifice through preparation and training. The goal of eliminating all actions that cause pain would lead to paralysis.

A possible way out of this pleasure-pain conundrum is to develop a utilitarian calculus that compares the pain an action causes with the pleasures it produces. Regrettably, the calculus of pleasure and pain fails to produce convincing equations because there is no common metric. We do not yet know how to equate the pleasure a hungry teenage boy takes from eating steak and eggs with the pain of a slaughtered steer and caged chickens. The calculus becomes even more convoluted when attempting to differentiate among types of pleasures and pains (see chapter 10). For example, to justify teaching a child to read, we must show that the unhappiness caused by turning off the TV and forcing the child to read is smaller than the future pleasures the child will gain through reading. That is, our utilitarian calculus must show that the intellectual pleasures of reading count more than the sensual pleasures of watching TV. To make such calculations work out in favor of literacy (and other socially desirable abilities), we are reduced to making assumptions about the relative worth of various types of pain and pleasure. The argument turns on rational analysis of what we think is right, not actual comparisons of painful and pleasurable experiences. As a result of these difficulties, pain and pleasure remain disputed criteria.

Is rationality a defensible criterion? When humans break a leg, they not only feel pain, but they also suffer; they dread the recovery time, mourn the lost opportunities, and regret the action that caused the injury. As far as we can tell, a rat merely feels the pain of a broken bone. According to this reasoning, the ability to rationally comprehend the implications of an injury is more important than the ability to simply feel pleasure and pain. Consider again the hungry boy eating steak and eggs. Only this time let's compare him to a hungry wolf eating a not-yet-dead deer. The boy is aware of his hunger and appreciative of it being satisfied. Perhaps this added awareness makes his pleasure worth more than that experienced by the wolf. Or perhaps the boy's actions are less defensible using this logic. Perhaps being aware that his diet causes pain and misery to other creatures obligates him to behave differently, beyond the cold calculus of pleasure and pain. Regardless of which argument we make, we have shifted the debate from pain and pleasure as the criteria to rationality as the criterion.

Unfortunately, rationality, language, self-awareness, sentience, and related criteria are problematic for at least three reasons. First, as chapter 10 on human nature explains, many animals possess these qualities to some degree, making it hard to draw the line that separates those possessing these qualities and deserving rights from those not deserving rights. Second, human infants, fetuses, and the insane do not possess these qualities, yet we grant them rights and protections. If we can't be consistent in our applications of rights, then we may be guilty of speciesism. Third, the emphasis on rationality may be arbitrary. What is it about rationality that makes it "good"? What makes rationality more deserving of special consideration than skin color? Merely to argue that rights should be granted to self-aware, speaking, conscious creatures is pure speciesism. It is not enough to say that these qualities are good simply because they are qualities that make us human. We also must defend why these qualities are deserving of special rights.

We forcibly separate infant chimpanzees from their mothers, cage them, and subject them to experimental surgery, even though they have some potential for language and self-awareness, more potential than some humans. But we don't subject any human, even those that have severely limited emotional and mental potentials, to the same treatments. What is our reasoning behind these decisions? Debates about these types of issues run aground on the rocky shoals discussed above—none of which were resolved here. Many debates end simply with people agreeing to disagree about fundamental issues such as the existence of rights, the equivalence of pleasures, the goodness of rationality, or the existence of a supernatural being granting humans special rights. The debate stops without finding common ground. Tempers flare and arguments get restated with greater passion but without hope of resolution. There are no easy answers, and we currently lack a conceptual framework to sort through the difficult issues.

WHAT RIGHTS DO WE GRANT NATURE?

The preceding section asked us to consider *what parts* of nature deserve rights and protections. Unfortunately, the difficult questions are not over. We must also ask ourselves *what rights* are deserved. Fortunately, answering this second question may clarify some of the issues raised by the first question.

One line of reasoning supportive of animal rights builds on the cherished democratic ideal of *equality*. The U.S. Constitution, for example, insists that all citizens are equal. But what does equality mean? Americans obviously differ from one another in our abilities; some of us run faster, some memorize better, and some sing more beautifully. We do not expect one another to have equal abilities; in fact, we celebrate our diversity. However, we do insist that we all have an equal *opportunity* to use our abilities and reach our potentials. When we seek to level the playing field, some people will still run faster, but all of us should be able to run in the directions of our choosing. Equality ensures people the opportunity to pursue and protect their interests. Focusing on these interests may help us navigate through some treacherous waters of bio-rights debates.[10]

Focusing on interests that matter also provides a way to avoid speciesist arguments. Charges of speciesism make us aware that the criteria used to assign rights need to be relevant to the rights being assigned. That is, rights should protect interests that matter to the unit of nature possessing the rights (rather than protecting something based on arbitrary criteria such as color of skin). These interests may provide logically defensible criteria for assigning rights if we can agree that some interests are logically more deserving of protection than other interests. We could then be logically consistent in decisions to sacrifice inferior interests to superior ones. But what are interests that matter and how do we decide which are superior and which are inferior?

Humans guarantee one another the right to pursue many interests, not just the interest in maintaining our own lives, but also our interests in happiness, dignity, health care, employment, voting, and free speech (see table 2, column 4). What interests matter to other units of nature? Bears, for example, don't have an interest in voting because they—arguably—cannot comprehend the human political system. They might, however, have an interest in being represented in a court of law by competent humans willing to defend bear habitat from bulldozers. Bears probably do have interests in eating, roaming, and mating. Perhaps they should have a right to pursue those interests. Dogs and trees do not have interests in making a last will and testament for disposal of their estates, but they might have an interest in benefiting from the will of a human who dedicates her estate to their welfare. Chickens might have interests in not living in the constrained, crowded, and accelerated breeding conditions of commercial chicken houses. A young calf might not need to roam the plains wild and free but might, arguably, have interests in not being starved, caged, and neutered for veal. An animal's interest in minimizing pain is often the most strongly argued point in animal rights debates. Pain seems an obvious interest because most animals react as if they care about pain—they avoid it in dramatic fashion.

Having interests that matter does not require possessing the conscious ability to recognize and defend those interests. For example, a creature exhibits an interest in living because it actively seeks nutrition, avoids threats, and procreates. Think of interests that matter as the behaviors and conditions a creature requires to survive and reach its potential. The contrast between basic, serious, and peripheral interests helps illustrate these points.[11]

1. *Basic interests* include food, water, and shelter.
2. *Serious interests* reflect things a creature can live without but not without difficulty or cost. For example, a bird can live without flight, a cow can live without pasture, and a human can live without literacy, but most all agree that the quality of such a life would be lower because the creatures' potentials would be thwarted.
3. *Peripheral interests* include comfort, convenience, status, and other opportunities that marginally increase quality of life.

Trade-offs among interests follow a relatively obvious and unambiguous logic within any one group of like things, such as a species. For example, a human's basic interests are the most important and should not be violated by any other human except those seeking to fulfill their own basic interests (i.e., the interest of self-preservation). People seeking to satisfy a serious interest, such as education, should not do so at the expense of another person's life or health. Peripheral interests such as wealth or status should be pursued only if they do not damage the basic or serious interests of others. Of course, there are exceptions, where people intentionally sacrifice their or another human's well-being, but we refer to these behaviors using value-laden terms such as altruism and exploitation to denote their moral significance.

This system for ordering interests does not overcome all difficulties, especially those of comparing conflicts among humans and other species, which is where we need the most help. It is difficult, for instance, to compare a human's serious interest in maintaining a cultural tradition of hunting and fishing with the basic interests of deer and fish life. Advocates of this method propose that we further categorize species according to their psychological capacity. This capacity includes the abilities to suffer pain and deprivations, be self-aware, experience diverse emotions, and engage in purposive actions. For example, an argument could be made that humans would hold the top category, but apes, elephants, dolphins, and some other mammals belong in the next category (see chapter 10, "Human Nature"). Insects, plants, and bacteria would be placed in much lower categories.

Those with greater psychological capacity can ignore some interests of species with lesser capacity. Thus, it would be morally justified for humans to violate the basic interests of a much lower species in order to satisfy a basic or serious interest of humans. It might not, however, be justified to do so to satisfy only a peripheral interest. Killing deer for food would be justified, for example, but killing for pleasure might not be. Similarly, starving and caging young steer to produce a desired taste and texture of meat (veal) would be inappropriate because taste is a peripheral interest. Animal experimentation for medical training and advances would be defensible as a serious or basic interest. Animal experimentation for cosmetics would not be justified because cosmetics satisfy peripheral interests. However, uprooting, pruning, injecting with chemicals, and tying tomatoes to stakes are morally defensible behaviors because vegetables have such a limited capacity to suffer. Thus, humans can abuse the basic interests of plants even to satisfy our peripheral interests.

The challenge with such a system, of course, is negotiating which species and which interests belong in which category. For example, as will be discussed below, hunters with long cultural traditions of family bonding, cultural heritage, and ecological education can argue that their interests in hunting are serious and not peripheral, just as subsistence communities can argue that hunting is a basic interest. This framework does not resolve the issues, but it does provide a means to structure negotiations in a way that may lead to consensus about some environmental policies.

BIOCENTRIC VERSUS ECOCENTRIC

The distinction between organisms and larger ecological units creates a critical fault line in discussions about bio-rights. Biocentrism focuses on the rights of individual organisms: you, me, Bambi, my dog, a maggot. Ecocentrism focuses on the rights of larger aggregate units of nature: communities, populations, species, ecosystems, and rivers. Biocentrism characterizes the concerns of animal rights activists, while ecocentrism characterizes the concerns of the Endangered Species Act. Heartfelt differences between biocentrists and ecocentrists can polarize environmental debate.

One of the reasons I've used the awkward language "units of nature" is to leave open the possibility that rights might be granted to more than organisms. It is often these larger units of nature that most concern us. We established policies and rhetoric stating our concerns about endangered species, biodiversity, wilderness areas, and the well-being of Mother Earth. Do we have more than instrumental concerns about these larger units of nature? There are many instrumental reasons to care, for example, about biodiversity; it provides aesthetic enjoyment, spiritual fulfillment, and a respectful reflection in the moral mirror. It may contain the cure to cancer, direct economic benefits, and indirect ecosystem services. But is biodiversity inherently valuable? Does it have interests? Do species have a *right* to exist?

These larger units of nature are problematic, in part, because they resist precise definition. Biological communities, species, and other ecological wholes have diffuse and overlapping if not arbitrary boundaries; they are not stable over time; and they otherwise lack the cohesion of an organism.[12] Ecosystems, biodiversity, and species, for example, are merely conceptual categories of ecological theory, concepts that help us understand and control nature. This is not to say that ecological and evolutionary processes do not exist independently of human observers. Rather, it is to admit that humans organize these processes into neat conceptual categories that we can describe, map, and manipulate. This argument has been presented in chapters 3 and 4, and will not be repeated here except to note that the alternative interpretations of ecological interactions (mechanistic, holistic, hierarchical) provide a case in point.[13]

In addition to being difficult to define, larger units of nature lack obvious interests. A species, for example, does not have an interest in persisting. Perhaps genes have an interest in replicating—that is what they do.

Organisms, similarly, seem to have an interest in avoiding injury. Species, however, don't act or do anything. A species is a conceptual unit developed by humans to categorize similar organisms into one category. Ecosystems, or ecological systems, are even more ambiguous. They are not bounded, finite, living things. They *do* organize matter and energy, and perhaps these nutrient cycles and energy flows could be interpreted as interests. But, as chapter 4 tried to illustrate, ecosystems do not strive toward an ideal state. They do not live and die; they merely change.

A river provides a good example of the difficulties defining the boundaries and interests of ecological systems. "Riverhood" has been proposed as an analogue to personhood and thus allowing rivers to possess rights and interests deserving protection in a court of law.[14] Defining a river, and thus defining what has rights, is problematic. Does a river begin at its banks, at the watershed boundaries, or does it extend to climate change and acidified rain? What are a river's interests? Is a river's ideal condition that which existed before, during, or after the last ice age when flow levels were dramatically different? Are plate tectonics harming the river's interests by tilting land masses and altering the rate and direction of river flow? Is it in a river's interest to have crashing, gurgling, roaring, and flooding waters? Would dams damage these interests? Answers to these questions are difficult and likely arbitrary, and quickly get mired in environmental fundamentalism. Opponents to damming might look back several thousand years and argue that the river's "natural" condition is free flowing. Proponents of damming might look back several hundred million years and argue that the river didn't even exist, or existed in very different form, so how can a dam matter? Arguments about what is natural typically mask people's own interests and values, not the river's interests. No one can win these arguments, as hopefully was illustrated in chapters 2, 3, and 4.

Ecological fascism is another difficult conceptual challenge to granting rights to larger units of nature. If we want to protect ecological wholes, such as ecosystems or species, then we must recognize that the rights of individual organisms will sometimes be sacrificed. The good of the larger system takes priority over individuals. For example, if a deer herd population explodes to the point that it overgrazes its forest habitat, eating all the new shoots of growth that enable tree regeneration and thus risking soil erosion and the long-term viability of the forest, then it might be necessary to control the deer population through hunting, relocation, or birth control, even if some of these actions violate the rights and interests of individual deer. Such logic assumes that the health of the ecological system is more important than the rights of the individual deer.

Management efforts removing invasive exotic species and escaped domestic animals have been charged with ecological fascism and further illustrate the tension between biocentric and ecocentric rights.[15] These "alien" species threaten an ecosystem's integrity because the exotics will eventually overwhelm and change species composition, creating a very different system. Land managers often find themselves in the difficult position of killing individuals so as to protect an ecological whole. For example, horses are exotic to North America, being introduced by Spanish explorers and nurtured by Native Americans and European settlers. Many horses escaped captivity and now run wild. These feral animals are routinely rounded up from arid lands in the southwestern United States because their grazing pressures are judged to adversely affect the health and integrity of the native ecosystem. For years these horses were merely shot, burned, and buried. Those near processing facilities were used for dog food and some, captured near human populations, were adopted as pets.

People taking a biocentric view of nature's rights objected to killing the horses. They disagreed that the horses' lives mattered less than ecosystem integrity. Some people merely objected to the inhumane shooting of horses when good homes might be found for them to live out their lives in locations not interfering with ecosystem integrity. Due to policy changes, many horses are now being relocated. But rough terrain and lack of adoption opportunities mean that horses are still being killed, often by marksmen in helicopters, in order to protect the integrity of a larger ecosystem whole.

Fascism is a political ideology that celebrates a nation or a race as some larger whole deserving loyalties and sacrifices from individuals. It subordinates all spheres of life to the ideological whole. Individual freedoms, liberties, and other rights get sacrificed in order to protect ecological wholes. If deer and horses can be sacrificed, must human liberties, and even human lives, also be sacrificed to protect the integrity of ecological systems? Logical solutions that prevent obviously unwanted fascist exploitation of individuals have been proposed. One such system prioritizes ethical obligations using four levels of importance.[16] The obligations in higher levels take precedence over obligations in the lower levels:

1. An obligation to respect the rights and needs of members of the immediate human community (i.e., families and associates);
2. An obligation to avoid suffering, famine, indignity, war, poverty in humans elsewhere on the globe;
3. An obligation to provide future generations of humans with functioning ecological systems, biodiversity, etc.; and
4. An obligation to respect the interests of other creatures on Earth.

This scheme is but one of many possible ways to trade off the rights of various units of nature. It does reflect many of the decisions we see in daily life. For each of us, there exists some hierarchy in the rights we grant others. We probably respect, in decreasing order of importance, the needs of family, neighbors, community members, fellow citizens, citizens of neighboring countries, and humanity elsewhere in the world. Only then do we see acts of concern directed toward nature. This particular scheme places the rights of functioning ecological systems above the rights of nonhuman organisms.

DIET, ANIMAL EXPERIMENTATION, HUNTING, AND OTHER MORAL DILEMMAS

Our lives are full of moral conflicts. Living our lives requires exploiting the lives of others. We repeatedly sacrifice the value and interest of nonhuman life. Public opinion polls suggest that many Americans are uncomfortable exploiting animals for the purposes of providing luxuries such as decorative fur and safe cosmetics. For many people, the peripheral interests of feeling better about appearances do not justify violating serious or basic interests by inflicting pain on animals. Using animals to develop and test medicines is another matter; here we have a conflict between the basic interests of humans and the basic interests of test animals. All but the most ardent biocenterist will consent to animal testing if no other testing options exist and the chance of saving human lives is great. Still, there exists ample room to debate when animal testing is necessary, whether the likely gains outweigh the pain, and how experiments can be carried out to minimize animal suffering.

Our diet provides a difficult moral dilemma. Most of us are content not thinking about the moral trade-offs we make and are happily ignorant of the food production system. Decisions are made for us long before the plastic-wrapped meat arrives on grocery store shelves. However, as the opening paragraphs of this chapter describe, animal suffering is an unfortunate side effect of the efficient factory-farm economy that provides us with a steady, cheap, and abundant supply of chicken and eggs. The conditions of pigs, cows, and other domestic livestock are not much better. Eating kills and everyone must eat; it is a basic interest. In all but the most extreme cases (i.e., eating nuts and berries fallen to the ground), eating violates the basic interest of some other life-form. But we do have choices about what we kill in order to eat and how it was treated while alive. There are vegetarian diets, free-range chickens, and organic family farms that treat livestock with dignity. We can examine our dietary choices using the framework of basic, serious, and peripheral interests presented above.

Reasons offered for being a vegetarian range from saving money to improving health. The reason relevant to this chapter is bio-rights. Vegetarians satisfy their basic interests in eating without violating the basic interests of higher-order organisms that feel pain (e.g., mammals, perhaps fish). Vegetarians do violate the basic interests of organisms without much psychological capacity (e.g., vegetables). Vegetarians may argue that the aesthetic preference for a taste of meat is a peripheral interest that does not merit violating the basic interests of living animals. Vegetarianism also respects ecocentric concerns. By eating lower on the food chain, they reduce the need for livestock and for other demands placed on the land. It takes bushels of corn, tankers of gasoline, acre-feet of water, and tons of topsoil to produce a diet rich in animal proteins. Rather than feeding those resources into livestock production and creating wastes of heat, pollution, and sediment, the corn and other crops could bypass meat and dairy production, going directly to a human diet. The same caloric and protein intake provided by a pound of beef could be achieved at a fraction of the environmental costs with a vegetarian diet.

However, vegetarian or organic diets are not clear-cut environmental winners. Eating locally grown, pasture-raised beef might be more ecologically benign than eating irrigated soybeans grown thousands of miles away on industrial farms. Grazing can be among the least damaging agricultural practices, while the irrigation, pesticides, fertilizers, and cross-country transportation required to grow genetically modified soybeans can have enormous environmental impacts. Many more wild creatures survive in pastureland and forests than in cornfields. Likewise, pesticides and transgenetically modified crops are not unambiguously good or bad. The increased crop productivity these non-organic technologies allow can reduce the need for pastures and plowing, thereby reducing the need to convert species-rich forests to pasture and erode soil through tilling. Thus, there are biocentric and ecocentric arguments for and against vegetarian and organic diets.

Hunting is often contested on grounds that it violates animal rights because it takes lives and causes pain. Sometimes hunting is justified as satisfying the basic interest of food. More often it is justified by the experiences it generates, the education it provides, and the control of wildlife population it affords.[17] Some argue that latter reasons are serious if not basic interests that justify taking the lives of hunted animals. Critics see hunting for anything but food as satisfying only the peripheral aesthetic interests and thus not justifying violating an animal's most basic interest. Hunting, they argue, is equivalent to torturing rabbits in tests of cosmetics because hunting implies the value of a life is less important than hunter entertainment. Some critics argue that hunting is even worse, suggesting that it cheapens the value of all life, even human life, by encouraging people to think that life can be traded for sport.

Proponents argue that hunting is much more than a sport. It provides profound aesthetic experiences unavailable elsewhere. Hunters experience deep relaxation and meditation. They often disregard food and sleep. They sit quietly for hours and must maintain a constant state of alertness as they attend to the sights and sounds of possible prey. This state of relaxed alertness echoes the unhurried, concentrated awareness characteristic of Zen-like meditative states. The intense rush of focused excitement that occurs at the moment of killing acutely contrasts with the meditative state, creating a unique experience.

Hunting also provides profound educational experiences that arguably are serious if not basic interests of hunters. Deep family traditions exist where sons, daughters, fathers, mothers, uncles, and cousins spend time together at hunting camps and on hunting trips. Stories are told, traditions nurtured, and families bonded. Hunters typically must be ecologically literate to succeed at hunting. Their field experience teaches them about the habitat and habits of their quarry. Hunting provides people a primary source of identity and purpose in life; they become deeply committed, highly specialized, and very serious about their hunting (or fishing). The friends they keep, the literature they read, the location of their residencies, and even their work all revolve around hunting. One duck hunter describes how his hunting dog was more important than his marriage:

> I've been married twice, and hunting's done me in each time.... Look what happened last time.... I had this old dog, he was a damn good dog, too, but by God if he didn't raise a little hell around the house once in a while. The old lady didn't go so much for that, you know, and she says to me one day, "Al," she says, "it's either me or that god dammed dog." Well I thought on that for a while, then I say to myself, by God, you can't find a good dog like that every day. No sir, that's the truth. So I say to her, "Why hell, that's no choice," and I helped her pack her bags. And she went, too, by God. Listen, I love to hunt, let me tell you. The only reason I work is so's I can buy guns, ammo, and dog food.[18]

Hunting sometimes provides the most practical means to control populations of species that adapt well to human habitats. Geese and deer are examples. Deer are now so plentiful in some suburban areas that they denude gardens and landscaping; they cause enormous property damage and threaten human life by causing car accidents and further threaten public health through disease vectors such as ticks. As a result, in some communities the positive image of Bambi is changing to the negative image of a hoofed rat.

Relocating troublesome deer is not only expensive, but many deer die in the process and few places exist where additional deer would be welcomed.

Birth control is also expensive and, thus far, ineffective. Hunting remains the most economically efficient way to reduce deer herds, reduce property damage, prevent herd starvation, and protect forest regeneration. Animal rights advocates often find hunting an unacceptable solution. Feeding programs provide an alternative to hunting, but these programs generally are not sustainable because well-fed herds continue to increase in size, necessitating more habitat and more feeding. Introducing "natural" predators such as wolves is another alternative, but few biocentric arguments are satisfied by trading the predator's teeth for the hunter's bullet. Workable solutions are elusive.

CONCLUSION

Rights provide a potent counterweight to instrumental values. If a unit of nature possesses rights that humans recognize and respect, then decisions to manipulate it for instrumental reasons must be carefully justified. Those units of nature possessing rights are protected, to some extent, from exploitation. Rights can be violated, but we must debate and provide good reasons for these decisions—just like we debate the death penalty and war. Units of nature without rights will be subjected to the calculus of instrumental values. Certainly some of these instrumental values, such as aesthetic appreciation and moral reflection, motivate preservation and humane treatment of organisms and other units of nature. But other instrumental values, such as economic profit, justify destruction and exploitation.

Deciding which units of nature merit rights is a perplexing and challenging task. We should avoid speciesism, just as we should avoid racism and sexism. One way to do this is focus our arguments on interests that matter to units of nature. Life, liberty, the pursuit of happiness, equity, and justice are interests relevant to Americans, if not all humans. Other units of nature have other interests. These interests, the units that possess them, and the methods to protect them are all subject to negotiation, testing, and refinement. We constantly debate, test, and refine what we mean by human interests and rights. Why should the rights and interests of other units of nature be any different?

Even if rights could be unambiguously assigned, it is not at all clear whether doing so would improve or hinder environmental negotiations. The polarization between economic and ecological worldviews, between humans and nature, might be accentuated by assigning rights to units of nature. Deciding which units of nature deserve rights does not dissolve the human-nature dichotomy that is the source of such polarization; rather it reinforces it. It separates those deserving special protections from those not deserving.

Most of the arguments in this chapter have focused on where to position the line separating those with rights from those without rights. A repositioned line just changes the fulcrum over which the human-nature dichotomy balances. Repositioning the line just reinforces our obsessive focus on the differences between humans (and those few others to which we extend the privileges of our community) and the rest of nature. Rather than drawing lines to exclude others, perhaps we should focus our energy on understanding the specialness present in all nature. Instead of focusing on the differences between those with and without rights, perhaps we should focus on the special qualities inherent in all life. We need to understand nature in ways that helps us appreciate, negotiate, and sustain all nature's special qualities. Stephen Budiansky reaches a similar conclusion after analyzing animal rights debates:

> An honest view of animal minds ought to lead us to a more profound respect for animals as unique beings in nature, worthy in their own right. The shallow and self-centered view that sees what is worthy in nature as that which resembles us seems vapid and petty by comparison. We try so hard to show that chimpanzees, or monkeys, or dogs, or cats, or rats, or chickens, or fish, or frogs are like us in their thoughts and feelings; in so doing we do nothing but denigrate what they really are. We define true intelligence and true feeling in human terms, and in so doing blind ourselves to the wonder of life's diversity that evolution has bequeathed earth. The intelligence that every species displays is wonderful enough in itself; it is folly and anthropomorphism of the worst kind to insist that to be truly wonderful it must be the same as ours.
>
> It is always dangerous to try to draw moral lessons from the blindly amoral process of evolution. But if there is a lesson at all here, it is that all of the creatures that evolution has fashioned are remarkable in their own right. All have hit upon unique ways to make a living against all probability. And that is something to respect, and to treasure.[19]

* 12 *
Aesthetic Nature

Parks and Recreation
Extraordinary Experiences • Beauty • Neatness

Americans spend many of their days and dollars renewing and improving themselves while relaxing in, learning from, and generally enjoying nature. The social and economic benefits are profound, well documented, and total in the billions of dollars.[1] Take mountain biking as an example. Its popularity skyrocketed in the last few decades, with races and events occurring nationwide. Ski resorts now have enough mountain-bike clientele to stay open during summer, and miles of hiking trails in public forests and parks have been redesigned to meet biker needs.

A survey by the International Mountain Biking Association found that its members average ninety-five rides each year, once every four days! These people are serious about their recreation. Biking is more than passing time or getting exercise. It's a way of life that provides esteem and identity. Bikers work at jobs that allow ample free time during daylight hours, and they buy houses near biking trails. They read literature about biking and spend their disposable income on biking equipment. Their friends and acquaintances are mountain bikers. Who they know, where they live, and what they do revolves around their recreation.[2]

These mountain bikers may seem extreme, but consider other popular recreation activities: bird-watching and gardening. Birding enthusiasts deprive themselves of sleep to rise early in the morning and commute long distances to remote locations where they then walk along cold, wet, rough trails. They stand motionless and silent for long stretches of time, straining through field glasses for a fleeting glimpse of shy birds. To some this behavior may sound like a form of punishment or torture, but to birders it is paradise. Birding is one of the fastest growing recreational activities, and it is big business: collectively, birders spend big bucks on bed, breakfast, birdseed, and binoculars.

Many Americans toil, sweat, and blister while planting, weeding, fertilizing, watering, and pruning their vegetable gardens. From a financial perspective, the behavior makes little sense. Industrialized agriculture provides ample foods at low prices conveniently transported to neighborhood stores. Most gardeners could make more money working a paying job than they save by not buying vegetables at the market. Gardeners quickly point out that their homegrown produce may be healthier and tastier as well as cheaper, but few will deny that part of their motivation is the experience of gardening. Gardening is their recreation: a means to escape and relax.

Avid pursuit of recreational experiences and aesthetic settings shape and have shaped the American landscape. Billions of visits are made each year to public parks, gardens, and other natural recreation areas. Millions of acres of public lands are developed and managed to provide recreation access for scenery and solitude. Whole government agencies and professional societies service the nation's recreation needs. Our environment reflects our obsession with tourism; it is punctuated with scenic vistas, parkways, and other tourist destinations. These developments are typically surrounded by gas stations, hotels, and restaurants that service the tourism market. Recreation destinations have a domino effect on the landscape. Anyone who has visited a gateway city to a popular national park knows this landscape: Gatlinburg at the Great Smoky Mountains, Estes Park at Rocky Mountain, Mariposa at Yosemite, and Gardiner at Yellowstone. Closer to home, Americans spend billions of dollars on lawn care and billions more on second homes near natural amenities. Treating nature as an aesthetic resource has changed our landscape and shapes the way many people think about nature.

NATURE AS A PLACE TO RECREATE

Nature-based recreation was not always an integral part of American culture and identity. The Industrial Revolution not only changed the way people worked; it created our modern understanding of leisure as the opposite of work. Industry required vast coordination among many workers, who had to show up at the same time and place to perform specific tasks upon which one another depended. The flexible and independent work schedules of agricultural lifestyles and cottage industries had to give way to structured workdays and workweeks. "Work" became a time, a place, and an activity. During the "workday," people left their homes and went "to work"; while there they "worked." Leisure became a time, place, and activity occurring while not at work.[3]

Americans debate the proper place of leisure in society, struggling over conflicting tensions. On the one hand, leisure is thought of as rejuvenating, healthy, and able to teach teamwork, leadership, and other socially redeeming values. On the other hand, it is considered frivolous, wasteful, and self-indulgent. The Puritan glorification of work created distaste for leisure and outdoor recreation, which supposedly taught questionable moral values. In 1803 the *Massachusetts Missionary* magazine advised readers that sports were not fit for Christians and that youth should not waste their time in outdoor leisure pursuits such as hunting and fishing. At that time, few people knew how to swim and fewer went to oceans, lakes, rivers, or other natural areas to recreate. Those with the skill and desire to camp were oddities or outcasts.

Public tastes and opinion slowly changed. Progressives such as Benjamin Franklin championed outdoor recreation, especially swimming, as noble and healthy pursuits. Organized outdoor recreation games such as croquet and tennis began in the 1860s. Bicycles became popular in the 1880s, and *Wheelman* became one of the first recreation-oriented publications, promoting bicycles, bicycle clubs, and bicycle recreation. The first golf course appeared in 1888. Travel books began appearing even earlier, in the 1820s, glorifying what we now call tourism by recounting the adventures of early western explorers. Within several decades, traveling art and lecture shows popularized the awesome beauty and power of nature. The once "harsh and barren" mountains slowly became "scenic and sublime." Niagara Falls, for example, changed from being portrayed as evil and overwhelming to being portrayed as inspiring and uplifting. Publications in the 1870s such as John Wesley Powell's account of harrowing adventure and the inspiring scenery of the Grand Canyon introduced readers to western natural wonders that became important sources of national pride and identity. Mark Twain's *Roughing It* enthralled the nation with spicy western characters and grandiose landscapes, while William Cullen Bryant popularized nature with images in his *Picturesque America*.[4]

Participation in more primitive, nature-based recreation activities blossomed in the 1870s. Census figures from that decade show that the country had urbanized as people migrated from rural farms to cities in search of jobs and amenities. An urbanized public looked back with nostalgia at their rural past and yearned for opportunities to reconnect with nature and their pioneer past. Rod and gun clubs organized trips to the out-of-doors. The pseudonymous Frank Forester published a barrage of articles and books promoting outdoor adventure and woodcraft as a means to pass on cultural traditions. These and similar publications became extremely popular, birthing a literary genre continuing to this day. Gradually during the 1800s, outdoor recreation transformed from something uncommon and sinister to something appreciated and promoted. Nature became a place to recreate, and stories about fish that got away became part of cultural lore.[5]

East Coast "pleasuring grounds" became crowded by the late 1800s, so wealthy, trendsetting recreationists seeking novel experiences began traveling west along recently opened railroads. The first national parks were created, in part, to meet this demand. Congress and the transportation industry were easily convinced that there existed both political gain and profit in designating and developing parks such as Yosemite, Yellowstone, and the like. The parks were specifically justified as "pleasuring grounds" and as sources of "national identity."

Not only were parks and recreation good for the elite and the economy; they were supposedly good for the nation. Fresh air and activity in natural areas promoted mental and physical health while connecting Americans to their glorious pioneering past. The National Park Service's Organic Act was signed into law in 1916, charging the agency to keep scenic, historic, natural, and wildlife qualities "unimpaired for the enjoyment of future generations." This dual mandate of protecting the environment and maximizing public enjoyment would create serious challenges for park managers who now struggle keeping parks from being loved to death by tourists.[6]

Post–World War II economic development increased disposable income and free time. Roads, cars, books, and maps increased park accessibility, while public and private advertising campaigns encouraged people to take advantage of these opportunities and go "see the USA in their Chevrolet."[7] Baby boomer families responded. Recreational visits to parks and forests skyrocketed; from 1950 to 1984, annual visits to lakes and other lands managed by the Army Corps of Engineers increased from 20 to 500 million, and visitation to National Park Service lands increased from 20 to 300 million. Similar increases occurred on Forest Service and other public lands. At the dawn of the twenty-first century, these federal lands received nearly 2 billion recreation visits each year. Many more visits occur on state and private recreation lands. Recent polls show that 94 percent of people sixteen years and older participated in some type of outdoor recreation activity at least once a year. The most popular activities are viewing/learning activities such as bird-watching and nature study (76%), walking or biking trails and roads (68%), and social activities such as outdoor family gatherings (68%).[8]

Baby boomers became a constituency that knew nature as an aesthetic resource. They created a political climate that increased funding for the National Park Service and changed the Forest Service's emphasis from timbering to recreational developments.[9] Congress responded to this new political climate with passage of the 1960 Multiple Use Sustained Yield Act, the 1964 Wilderness Act, as well as the 1968 National Wild and Scenic Rivers Act and National Trails System Act. Other congressional acts reflecting the nation's growing concerns about aesthetic nature include the Highway Beautification Act of 1965 and the 1969 National Environmental Policy Act, which explicitly requires all federal projects to consider impacts to "aesthetic amenities."

The federal government now controls approximately 761 million acres, 691 of which are available for recreation—that is over 30 percent of the continental United States. Many million more acres are managed for recreational use by state, local, and private organizations. Billions of dollars in revenue are generated. Birding, hunting, and fishing, for example, generate about $100 billion each year, a large portion of that comes from licenses and excise taxes and is used to support wildlife habitat restoration and management by state agencies. About $35 billion is spent annually on sportswear, bicycles, outboard motors, and other recreation equipment. Servicing outdoor recreation travelers with food, accommodations, and related services generates almost 1 million jobs. Nature-based recreation has become a vital part of America's culture and economy.[10]

EXTRAORDINARY EXPERIENCES

American society has a tradition of providing natural resource recreation opportunities without charge or with nominal charge. Park and forest recreation opportunities are subsidized by general tax revenue because they are assumed to be good for society. What makes these experiences so special? Why should taxpayers provide millions of dollars and millions of acres for recreation? Aren't convenient, marketable substitutes available in the built environment? Many people have argued that nature-based experiences are special and deserve government support because they cannot be duplicated by human technology and because they make essential contributions to American culture. They argue that recreating in nature is different than recreating at a shopping mall.

The scale and complexity of nature supposedly has the power to overwhelm us, creating an oceanic feeling of being a grain of sand on the beach of time. Standing at the edge of the Grand Canyon, for example, provokes humility as we sense our frailty and insignificance. One's attention is totally captured by these settings. The mind becomes absorbed with the experience. One loses a sense of purpose, a sense of time, a sense of self, and becomes connected to and at one with something larger.

Natural recreation activities provide an intensity of experience often not possible in the civilized world. Nature surrounds, involves all senses, and sustains experience in a way no painting, poem, or sculpture can. Art objects have frames, boundaries, and stand on pedestals that set them apart. Nature, in contrast, is penetrated and touched. One becomes immersed in a natural area as opposed to viewing it from the outside. You smell the soil on which you lie, feel the cool breeze against your skin, hear the leaves rustle in the air, and watch puffy clouds float above a swaying canopy. Nature intensifies and changes your feelings: comfort deepens into bliss, stress eases into calm, and interest bursts into excitement. Consciousness becomes dominated by feelings of freedom, wonder, and curiosity rather than the minutia and concerns of everyday life.

We escape from civilization and go to nature. We could escape the stress and strain of everyday life by staying home, tuning out, and altering our consciousness with TV, music, exercise, or drugs. But nature provides something more. It is a place one wants to be and where the mind remains active. Nature restores our physical and emotional energies. Restoration begins with clearing the head of details from workaday life, creating a cognitive and physical distance from civilization that allows us to fill our minds with thoughts that matter to us, not to others. Our attention is not assaulted with advertisements, e-mails, or mundane tasks, so we recharge our ability to concentrate. Nature stimulates our arousal: leaves shake, birds chirp, brooks babble. No matter where we look—from the panoramic vista to a handful of soil—we find infinite complexity and in it a source of arousal. The combination of increased arousal and lack of distraction is special; it gives us the energy and ability to direct attention toward issues that often go ignored: the big questions about life, career, relationships, and God. Extended stays out-of-doors allow us to reflect on life's priorities and possibilities and perhaps escape the hazards of an unexamined life. One returns to civilization not just recharged, but with a better understanding of self.[11]

Contrasts between nature and civilization teach profound lessons. Taken-for-granted luxuries such as cold drinks, hot showers, and other qualities of life become noticed and appreciated upon return to civilization. Material abundance and technological conveniences are striking in comparison to the scarcity and manual labor of surviving with what you can carry on your back. The tone and pace of civilized life seem shrill and excessive compared to the rhythm and tranquillity of nature. Advertisements and distractions contrast with a sense of purpose and clarity of thought. Obligations and schedules replace the peaceful opportunities to ask deeper questions and deepen personal relationships.

When we return to our homes, we are surrounded by mass-produced artifacts in a rapidly changing, technologically oriented, human-dominated world. The authenticity of nature is lacking. During most of our lives, it is rare that we experience the original of anything. Our tables are coated with plastic that looks like wood, the artwork on our walls is copied, and the music we listen to is digitized. An original, in contrast, maintains an uninterrupted connection to the creator. Through experience of the authentic, we transcend the here and now, and the forces of creation become palpable. The creator might be an artist, a craftsperson, an engineer, evolution, or a supernatural being.

Consider a proposal to mine minerals from underneath a small, pristine barrier island that is now and will again become a park. All life and dunes are to be removed. The valuable minerals will be exposed, extracted, and transported to markets. The mining company promises to restore the island, re-creating the dunes, replanting the flora, and restoring the fauna. Highly trained ecologists will supervise the project, and on-site managers will monitor and manipulate nature until wild processes resume. We are assured that the mining company will stick with the restoration effort until ecological experts visiting the restored island can find no evidence of disturbance.[12]

Assume for the sake of argument that sufficient restoration technology exists to make this scenario possible and that we can trust the mining company to make it so. After the mining, the island will appear the same and ecological process will be fully functioning. Will any loss of aesthetic experience be only temporary? Can the full aesthetic experience be restored? Your answer may depend upon whether you know that the mining and restoration occurred. Some visitors may find the experiences less valuable upon learning the island's history. It would no longer be authentic, its connection to the original severed and its ecological processes no longer wild. Other visitors may find their experiences enhanced, feeling in awe of the human will, creativity, and technology that accomplished the re-creation. Regardless of how you answer this rhetorical question, the example hopefully illustrates that authenticity has value to some people. An important part of managing natural resource recreation areas, therefore, is maintaining natural appearances, wild processes, and opportunities for authentic experiences.

Authenticity is a fickle master, however. Providing people access to natural areas so that they might experience authenticity requires some level of development and artifice such as trails, maps, and toilets. These developments, in turn, encourage more visits, which further degrade authenticity. Moreover, some visitors care much less about authentic, wild experiences than about being entertained by a novel setting and comforted by convenient services. The National Park Service is challenged by its 1916 Organic Act to find the appropriate mix of authentic and developed experiences that maximize public enjoyment. Its purpose is

> to conserve the scenery and the natural and historic objects and the wild life
> therein and to provide for the enjoyment of the same in such manner and by such
> means as will leave them unimpaired for the enjoyment of future generations.[13]

Many park visitors expect to be entertained. The first park managers, eager to justify the existence of their parks, obliged. Legend has it that early visitors to Yosemite National Park were amused by throwing things off the mile-high cliffs—bonfires at night and chickens during the day. Some parks now coddle visitors with hotels, golf courses, post offices, grocery stores, sewage-treatment plants, hot-water showers, pizza delivery, camp-fire talks, guided walks, and five-star restaurants. Visitors' desire to observe wildlife provides a dramatic example of the tension between maximizing visitor enjoyment, protecting nature, and providing authentic experiences. Even though official policy frowned on it, early Park Service management practices in parks such as Yellowstone brought bears and people together. Parking lots and bleachers were built near dumps, where visitors were sure to find these elusive and charismatic animals actively foraging in plain view.

For many years the main concern about bears was that they attracted more visitor attention than the geysers, waterfalls, and other scenery; only later did safety become an issue. In 1951 the Park Service handed out more than a million leaflets warning visitors about injury and death caused by bears, but Yellowstone still reported thirty-eight bear-caused injuries that year, mostly associated with feeding incidents. Park managers relocated or killed bears that sought human food in campsites or were aggressive with tourists. Dramatic changes to the bear-feeding policies resulted in the 1960s not just because of concerns about visitor safety but because of concerns about preserving natural conditions and providing authentic experiences. These concerns began to dominate policies with the Park Service's adop-tion of a "natural regulation" policy—a policy partially motivated by the assumption that nature knows best and it is wrong to modify nature for entertainment purposes.

Following this policy, garbage dumps inside the park that fed bears and entertained visitors were closed and the garbage generated by restaurants, lodges, campsite, and maintenance operations was trucked out of the park. One of the desired consequences of dump closure was dispersal of bears across the park, away from dangerous confrontations with humans. Another hope was to wean bears from unnatural foods and to wean tourists from observing unnatural wildlife behavior. Respected bear biologists worried that dump closures would stress the already frail bear population. The bears' habitat had been seriously curtailed by changes inside and outside the park during the almost one hundred years when the dumps became part of the bears' diet. Farms and fences outside the park restricted migration, and fire suppression inside the park eliminated many open fields where elk grazed and berries grew, thus restricting important sources of food for grizzlies. Without the dumps as a source of nutrition, biologists worried that the bears' health might be at stake. Unfortunately, research that monitored bears also was curtailed by this policy, in part because the tags (used for identification purposes) detracted from the natural scenic and aesthetic values of the park's wild landscape. Over eighty bears drawn by hunger and an acquired taste for unnatural foods were shot by park staff during the next two years.

Grizzlies survive in Yellowstone today without access to dumps, but the species remains threatened. The actual impacts of habitat restrictions and dump feeding remain unknown, although intensive research has resumed. Current visitors infrequently encounter grizzly bears, except those classified as a nuisance and dangerous, therefore denying a recreation experience many people once enjoyed.[14]

With natural and authentic experiences becoming scarce, there is increasing debate about which experiences should be supported on public natural lands. A variety of experiences are sought in natural areas, ranging from the simple pleasure of viewing tame wildlife to the profound inspiration of deep time and solitude. Are some experiences better or more appropriate than others? Is there something "wrong" with seeing wildlife at feeding stations even if the feeding is unnatural? Aldo Leopold was clear that not all nature-based recreation experiences are equally justified. He discounted experiences diluted by a "gadget industry" that "pads the bumps against nature-in-the-raw." He lamented government programs to increase roads, facilities, and visitation because they fragmented, civilized, and crowded wild nature.

The highest and best use of nature-based recreation areas, Leopold argued, is to promote a "perceptive faculty" that makes people sensitive, receptive, and appreciative of nature: "recreation development is a job not of building roads into lovely country, but of building receptivity into a still unlovely human mind." Pleasure, novelty, excitement, and even reconnecting with America's formative frontier era are all of secondary importance. According to this logic, recreation managers should minimize facilities and discourage behaviors that distract visitor attention from the land and attract it toward artifice that might be more entertaining. Entertainment can be obtained elsewhere, in shopping malls and Disney movies. Nature is special in its ability to promote a perceptive faculty.[15]

Another way of asking whether aesthetic experiences produced by natural areas are unique is to ask whether socially acceptable, inexpensive, and unnatural substitutes exist. Good literature, athletic competition, and fine cuisine provide alternative means to generate pleasure, excitement, and relaxation. However, psychological restoration, spiritual inspiration, authentic connections to our past, and other culturally significant experiences might require contact with nature. American culture is full of heroes such as Mark Twain, John Muir, Aldo Leopold, and Edward Abbey, whose descriptions and defense of these nature-based recreation experiences have inspired generations. Profound natural experiences seem solidly anchored in the cultural psyche. Opportunities for these profound experiences might be part of the cultural heritage we want to pass onto future generations.[16]

The dramatic vistas captured in coffee-table books and environmental club calendars define and promote picturesque qualities of nature. Most people experience nature visually, through the window of a touring car or framed on a postcard. Scenic parkways and scenic vistas dot the American countryside, and driving for pleasure is one of the top recreation activities. The Blue Ridge Parkway provides a famous example. Its two lanes were built exclusively to promote scenic tourism, traversing 470 miles of the southern Appalachian ridgetops and winding through a landscape meticulously designed for scenic views. Everything visual is controlled: signage, facilities, vegetation, and background vistas. The signage is subtle, made of natural materials. Its consistency from one end of the parkway to the other provides a unifying visual experience. A common palette of materials, colors, shapes, and proportions is used to coordinate fences, bridges, gutters, and other engineered structures. For example, only rustic split-rail fences are permitted, but several arrangements of the split rails are encouraged so as to provide some visual diversity within this unity of materials and techniques.

Vegetation is carefully arranged and maintained. Roadsides are planted in grass and regularly mowed to maintain a narrow but not confining corridor that focuses the eye toward vistas and other scenic attractions. The grass corridors periodically widen into meadows that provide visual relief and contrast. Vegetation abutting mown grass consists of aesthetically pleasing and intentionally planted collections of dark green rhododendrons, contrasting nicely with red maples and punctuated liberally by flowering dogwoods. The winding road evokes a sense of mystery, with the constant promise of new scenic wonders located just around the next corner. Vista-blocking trees are killed and removed so that grand vistas of tremendous distances and high scenic quality can be seen as cars round the curving roadway.

Encounters with an economic landscape are minimized. Only the most primitive agricultural practices are allowed on and adjacent to Park Service properties, mostly grazing and hay making. Modern buildings, silos, and farm machinery are forbidden, and parking, restaurants, campgrounds, and other facilities are carefully screened from view behind earthen berms and clumps of vegetation. The parkway is but a narrow ribbon of land stretching along the ridgetops, so whenever possible the Park Service has secured easements on adjacent properties to protect the rustic, agricultural appearance of its view. Landowners with these easements cannot build modern structures or divide their land into smaller housing lots. The completely engineered scenic experience is highly successful: the Blue Ridge Parkway is consistently the most heavily visited national park.

The "picturesque" scenery typical of the Blue Ridge Parkway reflects a popular but debated type of aesthetic experience often captured by paintings, still photography, and cinema. Artists of the eighteenth and nineteenth centuries were instructed to minimize messy foreground details and strategically compose the scene so as to draw the eye toward charismatic focal points. Special devices resembling a car's rearview mirror allowed painters to compose a picturesque view out of a messy and confusing natural landscape. Artists faced away from the landscape they were painting and directed the mirror backward. Careful aiming of the mirror helped them frame just those qualities considered picturesque and ignore the rest of nature.

In 1757 Edmund Burke published *A Philosophical Enquiry into the Origin of Our Ideas of the Sublime and Beautiful,* an essay that provides methods and a rationale for appreciating nature. Sublime scenery impresses the viewer with God's grandeur. It promotes humility, the feeling of insignificance relative to the power and glory of God's creation. Such landscapes have, according to the Burkean scheme, rugged cliffs, sharp and angular lines, towering mountains, and dark, foreboding gorges. Beautiful landscapes, in contrast, promote feelings of peace and tranquillity, a state to contemplate God's grace, with their gentle, curving, soothing features. Artists went looking for sublime and beautiful views. They found some and created others. Their efforts likely still influence our aesthetic tastes.

John Muir, the famed writer and explorer of romantic nature, tells a story of being visited by artists looking to record on canvas views of Yosemite. Muir took the artists to several locations that inspired him, places where towering cliffs and roaring waterfalls exhibited nature's grandeur and evoked human humility—landscapes Burke might call sublime. The artists wanted nothing of these landscapes and were not satisfied until Muir showed them views containing elements that Burke would classify as beautiful. They painted and displayed only these views in their efforts to promote public appreciation for nature, landscapes, and national parks. He describes their search for the perfect view to paint:

> The general expression of the scenery—rocky and savage—seemed sadly disappointing; and as they threaded the forest from ridge to ridge, eagerly scanning the landscapes as they were unfolded, they said: "All this is huge and sublime, but we see nothing as yet at all available for effective pictures. Art is long, and art is limited, you know; and here are foregrounds, middle-grounds, backgrounds, all alike; bare rock-waves, woods, groves, diminutive flecks of meadow, and strips

of glittering water." "Never mind," I replied, "only bide a wee, and I will show you something you will like."

At length, toward the end of the second day, the Sierra Crown began to come into view, and when we had fairly rounded the projecting headland ... the whole picture stood revealed in the flush of the alpenglow. Their enthusiasm was excited beyond bounds, and the more impulsive of the two, a young Scotchman, dashed ahead, shouting and gesticulating and tossing his arms in the air like a madman. Here, at last, was a typical alpine landscape.[17]

It remains unclear whether properties of a picturesque view are determined by nurture or nature. Some formal aesthetic theories suggest culture teaches us what is pretty, just like the view from a mirror framed the picturesque or Burke defined the sublime. Rational thought produced these classic definitions of beauty that many of us now find so appealing. Different schools of thought would and do create different definitions of beauty, so goes the argument. For example, consider popular landscape paintings hung in art galleries. Thousands of people pay tens of dollars each to see agrarian landscapes stylized by Monet, Van Gogh, and other respected artists. Van Gogh's *The Harvest* depicts fields cleared of forests and replaced with crops, fences, and occasional buildings. The tall, widely spaced, old-growth trees favored by the currently popular romantic-picturesque aesthetic are absent. Also consider classical aesthetics, which promoted symmetry and order over naturalness and wildness. Mountains were considered ugly warts that blemished Earth's symmetrical sphere. Only with the association of God in nature did rugged mountains become sublime. Today they are considered picturesque.[18]

Others argue that the aesthetic appreciation of nature is innate. Humans evolved in the savanna, so they necessarily prefer open vistas dotted with trees, from which to scout potential sources of food with protective cover from predators. Water and lush vegetation attract our attention and interest because they are associated with conditions that promote survival. We are predisposed by pressures of natural selection, so goes this argument, to attend to and take interest in certain landscape features because primitive humans who did so survived to pass on their genes and become our ancestors.[19]

Regardless of the nature-or-nurture argument, aesthetic concerns have motivated enormous nature conservation and preservation efforts. Much of our park system and land-regulation efforts emerged from efforts to protect aesthetic views and experiences, and scenic views justify locations of parkways and designations of tourist attractions. Scenery also increases property values. Some home owners so value scenic vistas that they sneak onto neighboring property and illegally cut trees that block valued vistas, not to be dissuaded by even the highest of fines.

Most of us live in urban or suburban environments. If we look out the nearest window, we are likely to see mowed, planted, and pruned nature. The neat, tidy, tended appearance is an established and deeply embedded aesthetic convention. Neatness reflects the owners' intentions. A sloppy, littered, eroded landscape with broken trees, graffiti, and burned-out buildings communicates a lack of responsibility and care. Cultivated fields, mown lawns, strong fences, and painted buildings, on the other hand, convey attention to, concern for, and pride in the landscape.[20]

Neatness and aesthetics shape the ubiquitous suburban lawn. The idea of having a lawn surrounding a house did not emerge until well after the Civil War. Only the wealthiest people had gardens. Most people, if they owned property at all, had dirt around their houses along with chickens and other farm animals. Variants of grass that could survive heat and foot traffic did not exist, nor was there an aesthetic norm to plant and maintain a lawn. The advent of golf courses changed that. The first American golf course was built in 1888 and by 1902 there were a thousand courses. The first American president to play golf was Woodrow Wilson (1913–21), and he instructed the U.S. Department of Agriculture to develop grass variants that would improve golf course design. Those same grasses made lawns practical. The wealthy and elite planted lawns as places to play golf, croquet, and tennis. The middle class then strove to mimic the style and status of the elite, and by 1949, just after World War II, there were 19 million lawns, a number that doubled by 1960. The chemical industry, looking for peacetime applications of wartime toxins, promoted chemical lawn-care products with aggressive advertising in home-and-garden magazines:

> It is time to take up arms against the weeds. From now on, when man and nature meet on the lawn, it's dog eat dog. . . . Your best bet is . . . wholesale slaughter by chemical warfare.[21]

Lawns are now ubiquitous, and unsightly lawns can produce fines from town officials and scorn from angry neighbors. The social and environmental impacts of imposing this aesthetic on our landscape are staggering. Yard wastes can account for 20 to 50 percent of landfill usage, and lawn maintenance can account for two-thirds of a municipality's water consumption. Some streams in suburban areas have two to four times the pollution of streams draining agricultural areas because of excess pesticide and fertilizer applied to lawns. In 1984 more synthetic fertilizer was used on U.S.

lawns than for all of the agriculture in India. The amount of money spent on lawn care in 1964, during the Vietnam War, exceeded the money spent on foreign military assistance. This comparison continues as annual U.S. expenditures on lawn care have increased to many tens of billions of dollars and now rival expenditures for wars in Iraq and Afghanistan.[22]

Enormous progress is being made by minimizing and composting yard wastes, planting grasses that require less water, and promoting alternative aesthetic expectations among home owners. But the classic suburban lawn that increasingly dominates our landscape still has tremendous impacts. Rural lands and wild forests are being converted into houses and lawns at an alarming rate. As housing lots fragment contiguous ecological systems, the ecological functioning of whole regions can change.

CONCLUSION

Arguably many U.S. national parks, forests, and wilderness areas exist because of aesthetics. The public supports land uses that produce scenery and recreation at least as much as they support land uses that protect biological and ecological qualities. Aesthetic landscapes attract attention; people get interested in nature. Through use and familiarity, this interest may turn to affection and eventually promote understanding of the landscape's broader values. The great works of human art not only attract attention but also sustain it. One returns again and again to the creations of Michelangelo, Pablo Picasso, and Frank Lloyd Wright, each time seeing more, learning more, and interpreting more from the cultural context and artistic traditions in which these artists were embedded. As reviewed in chapter 4, "Ecological Nature," Leopold argued that aesthetic appreciation of landscapes can lead to ecological, evolutionary, and anthropogenic appreciation of nature, which in turn may motivate a sustainable land-use ethic.[23]

* 13 *
Moral Nature

Thomas Jefferson
Henry David Thoreau • Environmental Education
Identity Transformation • Moral Mirror

What characteristics make communities thriving and sustainable? Previous chapters defined some of them. Certainly such communities sustain life and livelihood with continuous flows of ecosystem services and economic resources. But thriving communities are also beautiful, spiritual, healthful, equitable, moral, and create a sense of hope, pride, and identity. This chapter focuses on the last of these characteristics, which are among the most cherished and contested qualities of nature: the meanings and moral lessons people find in nature and use to define themselves and their communities.

We have a long history of looking to nature for sources of morality and identity. English Enlightenment philosopher John Locke (1632–1704), whose influence on America's founders is hard to overestimate, repeatedly invoked the moral authority of nature to justify his political positions.

To UNDERSTAND political power right, and derive it from its original, we must consider what state all men are *naturally* in, and that is a state of perfect freedom to order their actions and dispose of their possessions and persons as they think fit, within the bounds of the *law of nature*, without asking leave or depending upon the will of any other man.

A state also of equality, wherein all the power and jurisdiction is reciprocal, no one having more than another; there being nothing more evident than that *creatures of the same species and rank, promiscuously born to all the same advantages of nature and the use of the same faculties,* should also be equal one amongst another without subordination or subjection; unless the lord and master of them all should, by any manifest declaration of his will, set one above another, and confer on him by an evident and clear appointment an undoubted right to dominion and sovereignty.[1]

Thus, Locke used the moral authority of nature to justify freedom and equality of humans, two qualities that form the bedrock of American culture and governance. Indeed, the moral authority of nature is deeply embedded in much of our language and logic. Thomas Jefferson, Henry David Thoreau, and other pillars of American politics and culture relied on nature as justification and rationale.

THOMAS JEFFERSON, PASTORALISM, AND SUBURBIA

Thomas Jefferson (1743-1826) appealed to the authority of nature as justification for many of his positions. Supposedly it is difficult to find a page of his writing that does not use the words "nature," "natural," "Creator," or other synonyms.[2] Lessons, inspired by the Creator and taught by nature, he argued, held greater truths than lessons from history, tradition, law, politics, and other human authorities. Several of these natural laws provided Jefferson justification for his vision of the democratic experiment that was to become America. Note the opening sentences of the Declaration of Independence:

> When in the Course of human events, it becomes necessary for one people to dissolve the political bands which have connected them with another, and to assume among the powers of the earth, the separate and equal station to which the Laws of *Nature and of Nature's* God entitle them, a decent respect to the opinions of mankind requires that they should declare the causes which impel them to the separation.

In particular, Jefferson believed a pastoral lifestyle and landscape would ingrain a necessary moral fiber in Americans. Engaging nature through agricultural labors, he reasoned, built independent and self-reliant character. Pastoralists make what is needed, and perhaps just a bit more to trade for goods, services, and luxuries that the land does not provide. In contrast, an economy built on industry and service forces people to be dependent on others for life's necessities. Workers become subservient to the captains of industry and the "caprice of customers." They can't control their employment opportunities or their sense of self-worth that such control provides. They are closely supervised by self-serving employers, who are in turn beholden to whimsical and self-serving consumers. People who work their own land, in contrast, have both autonomy and dignity.[3]

Pastoralism also promotes sustainability. The pastoralist is committed locally and for the long term. Farmers won't relocate on a whim or in search of higher wages or lower taxes. They realize that today's management choices impact tomorrow's productivity. For example, seeding or fertilizing or leaving a pasture fallow makes the whole farm more productive years down the road. If there is no farm, then there is no current income and, just as important, no future income to support the farmer's retirement.

The pastoralist also is humble and the pastoral society egalitarian. Personal fortune is proportional to the amount of work invested. Unlike an industrial/service economy where owners of capital get rich from the labors of others, a pastoral economy rewards each according to their labors. A classless, egalitarian society results, without the accumulated wealth that allows a few to dominate the many. Nature can be harsh, making the pastoralist humble and cautious. Weather and insects can create or destroy a year's crop and a life's work. The cost of failed technologies can be catastrophic. New practices are evaluated with humility and caution against the tried-and-proven technologies of tradition.

Jefferson pulled no punches in this rhetorical defense of pastoralism. People not directly involved in land cultivation risked "subservience," "venality," and the sin of "ambition." Crowded European cities, where manufacturing dominated agriculture, were "sores" on the body politic, producing people with flawed moral characters that erode "laws and constitutions."

> Those who labour in the earth are the chosen people of God, if ever he had a chosen people, whose breasts he has made his peculiar deposit for substantial and genuine virtue. It is the focus in which he keeps alive that sacred fire, which otherwise might escape from the face of the earth. Corruption of morals in the mass of cultivators is a phaenomenon of which no age nor nation has furnished an example. [In contrast, urbanization and industrialization] . . . begets subservience and venality, suffocates the germ of virtue, and prepares fit tools for the designs of ambition. . . . The mobs of great cities add just so much to the support of pure government, as sores do to the strength of the human body. It is the manners and spirit of a people which preserve a republic in vigour. A degeneracy in these is a canker which soon eats to the heart of its laws and constitution.[4]

It is important to put Jefferson's position in context. He was arguing against Alexander Hamilton's vision of a stronger federal government that would function to promote finance, trade, and industrialization. Rather than build cities and manufacturing centers in the United States, Jefferson argued that raw materials be produced in North America and transported across the Atlantic for processing in Europe. The lost wages and cost of transportation would more than be made up by the absence of the urban pollution, vice, and societal malaise that industrialization and urbanization produced. Jefferson worried that an industrial economy and urbanization might produce corruption and class warfare that could destroy a struggling democracy.

Jefferson's personal life reflects the tensions between pastoralism and urbanism. As a public servant, he often lived and worked in cities, yet he repeatedly "retired" from political life to his country retreat at Monticello. As a U.S. ambassador, he enjoyed many years in great urban cites of Europe, where he developed a passion for architecture, furniture, literature, and food, importing many European fineries and ideas back to his mountaintop retreat in Virginia. No doubt many of us feel this same tension today: the yearning for simplicity and self-sufficiency tugging against the appeal of culture, convenience, and diversity.

Later in life, Jefferson came to support American industrialization. He realized that being dependent upon Europe for manufacturing made America vulnerable to blockades and tariffs. He rationalized that conditions unique to America could prevent the problems that manufacturing and urbanization caused in Europe. The abundant, fertile nature of the American frontier would always provide a relief valve where those who felt exploited by oppressive industrial labor practices could flee the urban-industrial labor force. Industry would be forced to provide safe, dignified working conditions in order to retain scarce labor. Meanwhile, people living in and learning from the frontier would infuse American culture with character and prevent industrialism from corrupting America.[5]

The shine on Jefferson's pastoral idealism tarnished when put to the test of reality. The noble yeomen, husbanding the land and shepherding a family, actually worked very long hours at grueling and demeaning tasks. Settlers and small farmers lived in poverty, with limited access to culture, education, or organized religion. The American economy, culture, and landscape evolved away from small family farms to become dominated by industrialization and technical innovation. Still, the pastoral ideal remains embedded in the American psyche. Current popular writers, such as poet-farmer Wendell Berry, continue to provide powerful and inspirational accounts of the need to interact with nature through agriculture. Evidence of the pastoral ideal can be found in popular movements such as Farm Aid to subsidize small family farmers against agribusiness, as well as in arguments for organic farming and against industry-intensive biotechnology, for bioregionalism and against the global economy, and for smart growth and against sprawl.

Suburbia is a derivation of pastoralism that emphasizes the aesthetic of living near nature rather than engaging nature through work.[6] Starting in the mid-1800s, scientific, professional, and popular literature encouraged people to abandon urban living and adopt a rural lifestyle. Books such as *Homes of American Authors* (1853) and *Homes of American Statesmen* (1854) popularized the status and enjoyment that country homes provided. In 1848 Andrew Jackson Downing wrote a column titled "Hints to Rural

Improvers." It is an early example of the very public debate motivating urban-to-rural migration:

> In the United States, nature and domestic [rural] life are better than society and the manners of towns. Hence all sensible men gladly escape, earlier or later, and partially or wholly, from the turmoil of the cities. Hence the dignity and value of country life is every day augmenting. And hence the enjoyment of landscape or ornamental gardening—which, when in pure taste, may properly be called a more refined kind of nature—is every day becoming more and more widely diffused.[7]

By the dawn of the twentieth century, the United States had become an urban culture; large cities dominated economics, politics, and residential location. Suburban developments and weekend homes sprouted on the rural landscape as people looked for relief from urban living. Those with wealth converted abandoned or unproductive farms into country estates. The middle class bought modest homes on postage-stamp yards, joined country clubs, and attended summer camps. Suburban developments were designed with winding roads, borrowed views, and tree-lined boulevards to create an impression of living in nature. Picturesque landscape paintings, not agriculture or ecology, guided suburban development patterns that continue to this day.

Suburban residents and other pastoralists do not get their hands dirty harvesting crops or grazing cattle; rather, they observe others doing this work. The crop grown on the mowed, manicured, suburban landscape is aesthetics not corn. It is now part of the American dream to own a freestanding house with a yard, fence, mailbox, and driveway in suburbia. The result is a far cry from Jefferson's pastoral ideal.

The cultural currents of Jeffersonian pastoralism continue to thrive as a wave of "new" pastoralists trade their business-suit blues for their blue-jean dreams.[8] These new pastoralists leapfrog over suburbia into exurbia, creating new owners and new neighbors of nature. Technologies such as telecommuting, flexible workweeks, and the information superhighway fuel the counterculture urban-to-rural migration. Increased wealth and 9/11 security concerns combined with the age-old desire to live near nature also create an enormous market for vacation homes and cabin retreats. These pastoral opportunities may be nurturing American character, but they also come at the cost of fragmenting ecological systems into five-, ten-, and fifty-acre housing lots.[9]

HENRY DAVID THOREAU

Fellow citizens of Concord, Massachusetts, were not surprised when, on July 4, 1845, Henry David Thoreau (1817-1862) abandoned the town's civilized comforts and went to live a simple, frugal, and independent life in a forest owned by friend and mentor Ralph Waldo Emerson. Few were impressed by his antics. Even to his close friends, Henry could be cranky, contrary, and sarcastic. He offended townspeople by questioning their accepted wisdoms with intentional acts of civil disobedience. He refused to join a church, not because he denied the existence of God but because he questioned church ritual, which he thought bordered on the superstitious. He resigned his public school teaching position, resenting classroom micromanagement by the school board and rejecting their policy that required corporal punishment. He was jailed for refusing to pay taxes because he didn't want to finance a government pursuing an unjust war against Mexico and the unjust enforcement of slavery. And, as a rule, he shunned regular work. Rather than devote time to building a career and nurturing a family, he spent most days tramping across the landscape wearing shapeless old-fashioned clothes, carrying nature specimens and Indian artifacts, and scribbling observations on paper. Townspeople were particularly furious when a campfire he built to cook lunch accidentally burned much of the town woodlot. When it became obvious that he and his fellow camper could not contain the fire, Thoreau climbed a hill to observe the resulting forest fire rather than run for help and warn the town.[10]

Thoreau was a loner. His hero was that of the solitary scholar pursuing truth at all costs. He graduated from Harvard College at barely twenty years of age, eight years before beginning his famous experiment in simple living. While well versed in and impressed by the classics, he suspected that traditional book learning obstructed deeper truths. He would look for those truths while living for over two years in a small cabin he built at the edge of Walden Pond. *Walden*, the book describing the motivations behind his experiment and the truths he found, was published seven drafts and seven years later. Its critique and vision of America continues to influence how we think of ourselves, our culture, and our relationship with nature.

Thoreau worried that economic, political, religious, scientific, and other modern institutions created self-perpetuating and biased myths about how people ought to live. Members of his community seemed so caught up in these institutions that they ignored the bigger issues determining the quality and sustainability of life and community. For example, Thoreau believed that the good life could not be reached by running faster on the economic treadmill. The true cost of consumerism is the time consumers must spend on the economic treadmill pursuing money instead of elsewhere pursuing a fulfilling and meaningful life: a new pair of pants may cost a day; a large house may cost a lifetime. He saw people's spirits being "crushed and smothered" chasing the so-called "necessity" of economic progress. He argued that one should live simply and deliberately. A simple life provides time to consider and appreciate the beauty and moral insights of nature and humanity.

> Men labor under a mistake. The better part of the man is soon plowed into the soil for compost. By a seeming fate, commonly called necessity, they are employed, as it says in the old book, laying up treasures which moth and rust will corrupt and thieves break through and steal. It is a fool's life, as they will find when they get to the end of it, if not before.... Most men, even in this comparatively free country, through mere ignorance and mistake, are so occupied with the factitious cares and superfluously coarse labors of life that its finer fruits cannot be plucked by them.... [The laboring man] has no time to be any thing but a machine.[11]

Thoreau trained himself to see aspects of life that normally went unnoticed. He dug deep, beyond cultural filters, into the bedrock of life to find the foundation of truth and meaning. He emptied his mind of conventional wisdom and sought truth through the primary experience of living a simple life: "My head is hands and feet," he explained.[12] He believed he would not find truth using classical concepts taught in schools and codified by conduct of civilized life, but rather he would find truth by living simply and with nature. He attended to the details of living and deliberated everything he experienced. His experiment revealed deep currents and a possible direction for the emerging American culture.

> I went to the woods because I wished to live deliberately, to front only the essential facts of life, and see if I could not learn what it had to teach, and not, when I came to die, discover that I had not lived. I did not wish to live what was not life, living is so dear; nor did I wish to practise resignation, unless it was quite necessary. I wanted to live deep and suck out all the marrow of life, to live so sturdily and Spartan-like as to put to rout all that was not life, to cut a broad swath and shave close, to drive life into a corner, and reduce it to its lowest terms.[13]

One oft-quoted lesson Thoreau learned from his studies is that wildness—in humans and in nature—is an essential characteristic of a thriving and sustainable culture: "In wildness is the salvation of the world." Exactly what he meant by wildness remains debated. Clearly wildness meant more to Thoreau than just the unpeopled, unmanaged wilderness areas that dominate contemporary discussions of environmental management. In part, he seemed to be arguing that wildness can and should exist in people as much as in nature. It is the basic primordial or primitive quality that can be restrained and weakened by civilization. Wildness must be nurtured by walking the countryside, climbing a mountain, hoeing a bean field, studying a melting snowbank, wading into a swamp, or committing acts of civil disobedience. When nurtured, wildness promotes vigor, inspiration, and strength of character. Thoreau worried that the comforts, conveniences, and concerns of a modern industrial lifestyle had weakened European culture and would threaten American culture.

Thoreau speculated that the Roman Empire realized greatness because its founders (Romulus and Remus) were suckled by a wild wolf. The connection to wildness gave Romans vigor and strength. He reasoned that too much civilization caused Rome to lose its wildness and vigor: "the children of the Empire were not suckled by the wolf," and the Roman culture suffered accordingly. He observed that American wildness invigorated European migrants who arrived here "effete, sterile, and moribund." He worried that destroying American wildness could similarly threaten American culture. The hope, strength, and salvation of American culture require nurturing wildness not only in the American landscape but also in Americans, a quality Thoreau nurtured in himself through study and meditation of nature in the everyday world.[14]

Thoreau thought wildness also increased awareness or sensitivity to humanity's evolutionary and ecological connections. Wildness accentuates our connections, dependency, and debt to nature. It bridges the human-nature dichotomy and represents the change inherent in nature and essential to human culture. Human culture evolves or dies just as other units of nature evolve or die. Wildness is not other than human—it is human. Humans are wild, shaped by the same forces of natural selection that shaped the rest of nature. Humans are wild nature grown self-conscious.[15]

Thoreau also made observations later echoed by Leopold. Humans are part of an interconnected, interdependent, ecological land community, and each individual should be regarded "as an inhabitant, or part and parcel of nature, rather than a member of society." Hoeing a bean field to clear out weeds attaches Thoreau "to the earth." Through hoeing he realizes that the "beans have results which are not harvested by me." They grow also for woodchucks, bears, birds, soil microbes, and for the beans themselves. Nor are the beans produced solely through Thoreau's labors. To get from seed to harvest requires more than Thoreau can provide by himself. It requires soil, rain, sun, genetic instruction, and countless ecological interconnections. Moreover, Thoreau's industry with beans connects him to Native Americans and hence anthropogenic nature. His hoe finds arrowheads and pottery shards, reminders of the long and continuing human management of ecological systems. He wants us to think of beans as more than economic crops that produce profit but also as a connection to deeper meanings from which spring identity, hope, purpose, and other qualities of thriving communities.[16]

Thoreau died in 1862 of tuberculosis, struggling to his last moment to finish a book on the Maine woods. His work was only beginning to be appreciated at that time, but today he is credited with being perhaps the greatest American nature writer, the first truly American philosopher, and one of America's most insightful cultural critics. He used the lessons learned from his experiment to critique modernism and the economic worldview. This critique provides departure points for a wide range of environmental ethics, including Muir's questioning of Judeo-Christian anthropocentrism, Leopold's ecological land ethic, bioregionalism, ecofeminism, and deep ecology. Thoreau's insights about interconnectedness predated ecology and influenced Leopold, and his grasp of evolution predated Darwin. His willingness to practice civil disobedience in defense of morality influenced both Mahatma Gandhi and Martin Luther King Jr.; and his appreciation of human creativity and its connection to nature was revolutionary. Thoreau is the fountainhead of American environmental thought, and his pursuit of truth and meaning through living simply with nature helped establish a tradition of looking to nature for lessons and insights about how humans should treat one another and the world around them.

ENVIRONMENTAL EDUCATION

In 1893 Frederick Jackson Turner published *The Frontier in American History*, in which he gave additional words and logic to the belief that wildness and the frontier played crucial roles shaping not just American policies but also the character of its citizens. American culture, he reasoned, grew from and was nurtured by the ever-present frontier. The frontier loomed large in the American mind: as a safety valve by dispersing social tensions, providing an endless supply of resources, a source of wildness, and a place to test and build character. The frontier symbolized economic opportunity, religious freedom, and relief from oppression.

American character supposedly was forged in the frontier by harsh weather, wild animals, Native Americans, and other hostile forces. The many challenges of survival produced people of strong and durable character with the common sense needed to guide a nation. Moreover, the continued promise of greener pastures and fertile farms forged a forward-looking society. Rather than looking backward, toward customs and technologies that worked in the past, Americans lunged headfirst into the frontier and into the future, confident that a little ingenuity and a lot of brute force could solve most any problem.

Americans celebrate heroes such as Daniel Boone and Davy Crockett, who helped settle the continent by blazing trails, guiding settlers, collecting information, and negotiating/fighting with Native Americans. They were the essential first wave of culture that helped America realize its Manifest Destiny, settling and defending outposts of Eurocentric culture. They cleared wilderness, plowed fields, and built towns. However, they also were impatient with the formalities and rituals of civilization; they fled to the frontier in search of absolute freedom and the wholesome spirit of wild nature. Daniel Boone, for instance, was frequently portrayed as someone who rebelled against the riches and status a founding father of Kentucky might enjoy; instead, he retreated into the frontier as civilization encroached. Myths have him relocating whenever people settled within a hundred miles of his cabin or if he could no longer fell a tree for fuel and have it land within a few yards of his front door. These frontier heroes survived using their physical prowess, practical cunning, ferocity, and courage. They were deemed morally decent because they communed with nature and practiced a simple moral code. Simple but not savage, they were children of nature weaned on a humble lifestyle uncorrupted by civilization.[17]

But by end of the 1800s, the frontier was conquered and the American landscape was mostly settled with farmers and merchants. In 1800 only 5 million people lived in what is now the continental United States, but by 1900 the population had risen to 76 million, leaving only thirty acres per person. Not only was the frontier symbolically closed around 1890 with the completion of the transcontinental railroad, but it was controlled by powerful individuals and business corporations, raising concerns about oppressive business and class interests. A psychic crisis resulted. People romanticized and mourned the nature, wildness, and frontier now lost and worried about the effects these losses would have on American culture.[18]

Educational policies in the late 1800s and early 1900s responded to concerns that "city children" were losing the "earnest, self-reliant and high minded" qualities forged by the frontier and essential to the American character. The generations of children raised in the concrete jungle might "become the very opposites of their fore-fathers." It was feared that they would remain immature and end up roaming city streets in primitive and warlike gangs.

Studies were commissioned to assess these dangers. Tests in the 1880s revealed that city children were ignorant of the names of most plants and animals, knew little of how food was produced, and knew nothing of the origins of leather or cotton. Then, as now, it was feared that the disconnection between people and nature would lead citizens to lose respect for the agrarian ideals and ecological systems on which the nation depended. Exposing children to nature became a priority so that future generations would understand their connections to the natural world and learn that butter did not come from buttercups. Textbooks about nature became part of elementary education, and after-school nature clubs provided outdoor activities. An association started by a high school principal in 1875 and named after the famous naturalist Louis Agassiz grew quickly to nearly a thousand local groups by 1887.[19]

Classroom learning alone was deemed insufficient. Children also needed opportunities to spontaneously explore nature, mimic animals, engage in tribal dances and games, and observe nature's cycles and rhythms. The absence of these experiences supposedly stunted city children's moral, emotional, and physical development. Nature study and exploration, it was hoped, would inspire "the aesthetic, the imaginative, and the spiritual in the [urban] child." At the dawn of the twentieth century, curricula guides such as *How to Study Nature in Elementary Schools* sought not only to teach the facts of nature but to expose children to nature and help them find the "goodness and beauty in nature so as to sweeten life and enlarge thought."[20]

If nature could not be found in cities, children must be transported to nature. "Fresh Air" charities organized and funded short escapes for urban poor. In the 1890s these programs helped over half a million children each year enjoy a day or two of hiking and swimming; thousands more spent nearly a week living with country families. The well-to-do could afford boarding schools that specialized in nature-based experiences. Summer camps were almost unknown in 1900, but by 1915 there were three hundred to choose from. Boys enrolled in the Snyder Outdoor School, for example, wintered on a Gulf island off Florida's coast while enjoying spring and fall in the Blue Ridge Mountains. Students devoted their mornings to more traditional lessons while spending afternoons immersed in hunting, fishing, mountain climbing, and sailing. By 1929 over 1 million children were summering at approximately seven thousand camps, which offered everything from business education and military training to woodcraft and wilderness adventure.[21]

Woodcraft Indians, Campfire Club, Boy Brigade, Sons of Daniel Boone, Girl Guides, Girl Pioneers, Girl Scouts, and other clubs sprung up to promote nature study, moral development, and public health. The Boy Scouts of America, perhaps the most familiar, consolidated many of the popular groups. It was founded in 1910 with an impressive national advisory council that included, among other notables, Gifford Pinchot and President Theodore Roosevelt. Early promotional material for the Scouts echoed the public health concerns and the restorative qualities of nature: "Boys in our modern life, and especially in our cities and villages, do not have the chance, as did the boys of the past . . . to become strong, self-reliant, resourceful and helpful, and to get acquainted with nature and outdoor life." Over 110 million boys have joined the Scouts since Congress granted it a charter in 1916, and the *Boy Scout Handbook* remains one of the bestselling books of all time. Uniforms, equipment, a literary genre describing scouting adventures, and a host of other consumables are eagerly provided by businesses able to cash in on the movement's popularity. The magazine *Boys' Life* typically has over 1 million subscribers. The Girl Scouts have a similar scope, reaching over 50 million females since the first troop organized in 1912.[22]

Environmental education at the turn of the nineteenth century mixed science, aesthetics, animal welfare, and progressive conservation concerns about exhausting finite resources. The movement's popularity declined in the 1920s but has reemerged in recent decades with a similar passion to motivate an appreciation of nature. Prepared lesson plans and prepackaged class exercises can be found in products with names such as "Project Learning Tree," "Project Wet," and "Project Wild," which emphasize, respectively, forest ecology, hydrology, and wildlife. Summer camps, scouting, and other organizations remain popular to this day. Some states, such as Wisconsin and Indiana, require environmental education programs in their public school systems.

Environmental educators continue to advocate for "ecological literacy" that uses environmental science to reinterpret traditional disciplinary and professional perspectives, including economics, business, engineering, sociology, and philosophy. The environment, it is argued, is more complex than traditional disciplinary perspectives can comprehend and more fundamental than economics and philosophy admit. Some of its most vocal and articulate advocates argue that environmental education should not be limited to factual content, but should also instill caring. Echoing the concerns advanced a century earlier, these modern environmental education advocates worry that the citizenry is not only uninformed about nature but uninspired by it. They argue that education has become "solely an indoor activity," that modern, urban, technologically dominated society has cut us off from nature and as a result we have accepted "ugliness" and "ticky-tacky" as the norm for the form of our communities. Children need to get into nature so that they can be inspired by its glory and power and therefore be motivated to protect and care for their environmental heritage.[23]

Recent calls for more environmental education have not been universally accepted. Some school boards around the country have debated banning textbooks that advance environmental protectionist themes. Residents in Escambia County, Florida, for example, sought to ban a high school science book they thought presented environmental problems such as global warming as undisputed facts as well as having an anti-industry bias. The Arizona state legislature overturned a 1990 law requiring environmental education in public schools. A middle school camp in Bend, Oregon, was canceled because its environmental themes were viewed as anti-Christian. In Idaho, state biologists were forbidden to discuss wolf reintroduction issues during their visits to public schools. Conflicts abound and tensions mount. Advocacy materials from environmental groups, public environmental management agencies, business, and industry inundate teachers. The environmental education controversy will likely continue for some time as powerful institutions of American life vie for the hearts and minds of the next generation.[24]

SOCIAL AND PERSONAL TRANSFORMATIONS

Understandings of nature have transformed our worldviews. Copernicus's observations of the night sky inspired theories about orbiting planets and demoted humans from the center of the universe; Newton's observations and explanation of falling objects suggested humans could predict and control nature by understanding natural laws rather than offer sacrifice and prayer to seemingly whimsical supernatural forces; Darwin's observations at Galapagos inspired theories about adaptation and evolution, demoted us from the pinnacle of creation, and embedded us in nature; and Einstein's observations of relativity inspired theories of quantum physics and placed us in an uncertain and chaotic universe that we cannot know objectively: these lessons from nature transformed human society.

Nature also transforms individuals. Camping, hiking, hunting, and fishing, for example, teach moral restraint, a quality some argue is lacking in today's self-indulgent society. The solitude of wild places frees people from laws, rules, and social customs that typically define and enforce "proper" behavior. People behave according to their own moral codes rather than peer pressure and learn about themselves through the consequences of their actions. A hunter, for example, can impose self-restraint and choose to never take a shot that might miss or wound the animal. The only acceptable shot is one that will, with all possible certainty, kill the animal instantly and with minimal pain. These ethically motivated hunters consider it irresponsible to shoot at fleeing animals, at great distances, or through thick foliage.

Nonhunters can practice moral restraint by minimizing their ecological impacts while hiking or camping. "Leave only footprints, take only memories" is a popular saying of the Leave No Trace organization, which promotes ethical recreation. Campers pack out trash and wastes, do not disturb or feed animals, minimize trail width to eliminate erosion, collect only fallen wood for fires, and minimize contrasts with the silence and greenery of nature by not singing around campfires or attracting attention with bright colors. Those unwilling or unable to venture off city sidewalks may still practice moral restraint when they encounter nature. Ants and worms can be sidestepped, leaves and flowers left attached to branches, and stones can be left unthrown at rivers and trees. Each of these choices builds moral character by having us practice ethical restraint. Cumulatively these choices can be transformative to individuals and beneficial to society. Presumably the moral character developed through interactions with nature will transfer to everyday behavior and benefit the community.[25]

The experience of wild places also provides rare opportunities to practice control over emotions. Recreationists must control their fears and anxieties when confronted with vertigo-causing cliffs, howling winds, or other extreme conditions where panic could lead to injury or death. Rarely in today's controlled and developed settings are people threatened with becoming lost, wet, or cold. Help is available at the corner store; at worst, it is a phone call away. But wild places do not have cellular phone towers, taxies, 24/7 convenience stores, or police. One must confront and control fear and anxiety. Likewise, when confronted with physically demanding tasks, one must resist the urge to quit and turn back. Only through perseverance and hard work, one step at a time, can a steep hill be climbed. Thus, American society values natural places because they teach self-reliance, self-control, and discipline. These lessons transform individual lives as well as the communities in which people live.

Last, but certainly not least, interests tweaked by nature can transform hopes, dreams, and behaviors. The sights and sounds of nature inspire awe, delight, and wonder, which can motivate lifelong passions. The morning chirps and fluttering colors of neighborhood birds may provoke interests that inspire a fulfilling career as an ecologist developing conservation strategies for endangered wildlife. A recreational experience may blossom into a lifelong hobby of greater and greater specialization and deeper and deeper meaning. Pleasant and profound encounters with nature may motivate people to transform their behaviors and live less environmentally destructive lifestyles. One may choose to turn off unused lights, recycle cans and bottles, take shorter showers, make fewer status purchases, and otherwise attempt to live lighter on the land and thereby protect the aspects of nature they respect and value. Such transformations are hoped for and promised by many of the experiential environmental education programs reviewed in the previous section.[26]

MORAL MIRROR

Our landscapes advertise our values. A yard strewn with litter, dead shrubs, and uncut grass suggests the owners don't care about their neighborhood and, perhaps, don't care about themselves. What is the message you get when you see some animal, person, or society foul its nest? What does it say about their priorities and concerns? We learn about ourselves by observing the consequences of our actions. Our actions toward nature are evident in the landscape for all to see. As America was being settled, a stump was a sign of progress—it showed that civilization was taking root. Later, industrialization, roads, and power lines symbolized accomplishments of which the nation was proud. For some, wilderness, clean air, and biodiversity are the symbols we should now promote.

Our landscape reflects and instills national values. It cements and projects our history. For these reasons, we build and maintain national monuments such as the Washington Monument, the Empire State Building, and the Golden Gate Bridge as symbols to remind us and others of our cultural values and achievements. Preserving nature and wild experiences, therefore, may be in the national interest because they symbolize values and ideals central to national character.[27]

The environmental conditions we leave behind shape the values, understandings, and expectations of future generations because people learn from their surroundings. If our descendants know only suburban lawns, shopping mall water features, vast industrialized agricultural fields, highway median strips, and ski resorts' developed mountains, then the nature they expect and value will be the neat and controllable settings that provide food and scenic beauty. If future generations know wild and dynamic nature, then they may be more humble in application of their will and technology. If they know a diverse, vigorous, and autonomous nature, then they may appreciate and value nonhuman life. But if they know only a polluted, crowded, built environment, then they will value only what they can clean and dominate: "If we leave them an environment that is fit for pigs, they will be like pigs; their tastes will adapt to their conditions."[28]

Wilderness areas, salmon runs, and biodiversity illustrate different values than do skyscrapers, interstates, and industrial agriculture. We shape the future by the lessons we leave behind. Which natures should we leave behind? Which lessons do we want our landscapes to teach? Which ideals do we want to sustain?

CONCLUSION

Nature's moral lessons contribute pride, identity, and direction to thriving and sustainable communities. But, as we have seen thus far in the book, there are many natures and hence many lessons. We should be cautious in advocating or accepting a single or fixed interpretation. We have a tendency to find in nature the lessons we go looking for. Locke found equality; Jefferson, integrity; Thoreau, vigor. Obviously, some of these lessons hold considerable merit for American culture. However, the merit of these ideals comes not from their naturalness but from their resonance with our own values and dreams. The abuses of social Darwinism should be remembered whenever lessons of nature are promoted as absolute truths.

History is littered with examples of people looking to nature for meaning. Nature is both symbol and mirror, pointing the way and reflecting the past. When we look to nature for a moral compass, we may instead find a reflection of our own values. Rather than an impartial guide, we find something closer to a blank slate onto which we write our own meanings. Rather than a crystal ball, we find muddied reflections of our assumptions and worldviews. Perhaps no other reason for valuing nature is more eagerly sought or hotly debated.[29]

Admitting that there are many lessons, and that many of these are socially constructed, does not mean that nature is meaningless or that the meanings in nature are unimportant. In fact, I argue just the opposite. The meanings we assign and ascribe to nature provide some of the most important reasons for valuing natural conditions. Lessons from evolution, ecology, anthropology, and multiple natures are essential to our search for sustainable and thriving communities. They promote an appreciation of our relationship and responsibility to nature. Lessons promoting equity, integrity, vigor, and a respect for nonhuman life add critical qualities to thriving communities. Deciding which lessons and which values to promote should be the focus of raging public debate.

Evolution, ecology, and anthropology do provide a few objective lessons, and we should adjust our worldview to reflect insights offered by these sciences. But, by and large, we should not seek cultural ideals from nature. Instead, we should place them there. Nature is a powerful symbol and mirror that can reinforce and reflect cherished cultural ideals. Constantly negotiating which ideals to promote through nature will keep nature front and center in our definition and construction of the thriving and sustainable communities we seek. Many of us want our children to wonder at the towering canopy of old-growth forests and future generations to appreciate the ecological interconnections that embed humans in the web of life. We think our grandchildren should witness the remarkable biodiversity made possible by continuing processes of natural selection. We believe future communities would be better if they could observe and learn to respect the long historical relationship by which humans shaped nature and nature shaped humans. We want people to have integrity, to feel vigorous, and treat one another equitably. We value these lessons and ideals because they promote the types of communities where we want to live, and we want to sustain qualities of nature that promote these values because we assume that future generations would be better off if they learned similar lessons and had similar experiences.

Conclusion

As the saying goes: We live in interesting times. Globalization and fundamentalism seem locked in a death struggle to control world economies and cultures. The biosphere, the thin skin of life that blankets Earth, is now dominated by human industry. Environmental alarmists look at this domination and see biodiversity loss, a destabilized climate, eroding soils, and overfished oceans. Environmental optimists, in contrast, see a reliable supply of food, energy, and other resources and control of the most serious air and water pollution problems. Meanwhile, the traditional methods of environmental management are faltering. Rational, centralized environmental planning is an admitted failure in most professional circles, and science wars have diminished the credibility of all expertise. Environmental issues infrequently find space on the national agenda, and critics say environmentalism's method and focus must change. These conflicting currents and eddies flow within the larger river of postmodern angst, causing us to rethink answers to our ultimate questions: What does it mean to be human? What is the essence of the natural and supernatural world we live in? How should we relate to that world?

We are indeed living in a time of transition. And it is a good thing, too, because we have difficult questions to ask and answer. Human civilization is facing unprecedented environmental challenges caused by the increasing materialist demands of an increasing human population. Some of the most serious challenges include soil degradation and erosion, global climate change, depleted fisheries, untreated toxic wastes, fragmented ecological functions, and declining biodiversity. Pluralizing nature may offer alternative ways to think about some of these issues. The challenge of biodiversity loss provides an example.

Hundreds of billions of people will look back on us with profound gratitude, visceral disgust, or sad empathy. A huge proportion of Earth's biodiversity will be lost or conserved during our lifetimes because of the actions we take. Extinction is permanent; most other environmental changes are reversible: breached dams can produce free-flowing river ecosystems, old-growth forests can reclaim farmland, ozone concentrations can thicken and again block life-threatening radiation, even bedrock exposed by coal mining can eventually sprout life-sustaining soil. But a species, once lost, is gone forever; the exact mutations of genes and tests of natural selection that produced it will not be repeated.

The current rate of extinction is one of the fastest in all of Earth's history—a rate comparable to changes that vanquished the mighty dinosaurs 65 million years ago. This time around, however, humans are causing the extinctions—and we know it. As we restructure ecological systems to satisfy our needs for food, shelter, income, and entertainment, we also remove and pollute the habitat of other species. The rate of these changes and the extinctions they cause show no signs of decline.

Why, then, with so much at stake, do we not unite in action to prevent mass extinction, or at least explicitly choose which extinctions we mournfully allow to occur? We know that species are going extinct. We know we are the cause. We even have viable solutions that can prevent some if not many of these irreversible events. Why don't we act? I suggest five primary reasons following from recent books on social collapse and environmental catastrophe by Jared Diamond and Richard Posner.[1]

- a distrust of claims that extinction is occurring;
- a pessimistic sense that human greed and population explosion are unstoppable forces that will overwhelm the best intentions and the most enlightened policies;
- a trust that technology will provide substitutes for losses that extinctions cause;
- a focus on spiritual life in the hereafter rather than on life and life-forms on Earth;
- and, most importantly, a lack of will to act.

Distrust Claims: Environmental alarmists and wise-use rhetoricians spin facts and figures for political purposes, making it difficult to separate fact from ideology and to balance benefits with costs. Decision makers are understandably confused and willing to delay action awaiting better science, which will always disappoint because scientific knowledge is based in probability while decision makers want certainty. Despite these obstacles, people willing to seriously evaluate environmental history will accept that biodiversity decline is occurring on our watch and that the causes of the decline are accelerating. Denial of this trend is a head-in-the-sand philosophy that dooms us to failure because humanity cannot solve problems it ignores. There is plenty of room to debate the exact rate, cause, and types of changes to biodiversity that are occurring, as well as the implications of these changes, but it is hard to deny that dramatic changes abound.

One of the profound challenges to mounting a response to biodiversity loss is that, compared to many environmental challenges, biodiversity is hard to see and the consequences of biodiversity loss are hard to comprehend. Sewage dripping from a pipe into our water or industrial emissions belching from smokestacks into our air can be seen, smelled, and tasted. Problems caused by these emissions are more immediate and more obvious—they affect human health and degrade human communities. Solutions seem within reach: plug the pipe, scrub the emission, or move your family away from the polluted area. But biodiversity loss is different: we rely on experts to tell us what biodiversity is and where it is decreasing. The consequences of biodiversity loss to our health and communities are less obvious and perhaps unknown, and corrective actions seem beyond the control of everyday people trying to survive the struggles of day-to-day life.

Yet our space-age computer technology and environmental science allow us to see our finite Earth from space, to comprehend the magnitude of change occurring in the biosphere, and to observe the interconnectedness of all life. We ignore these lessons at our peril. Anthropology and archaeology show how environmental degradation ended many previous human civilizations. We do not want to suffer the same fate. Apathy is not an option when the consequences are permanent and the risks great. This line of reasoning does not lead to the conclusion that all losses of biodiversity must cease. Rather, it suggests that we should condone only those losses we can live with.

Pessimism: The second reason for inaction is equally perverse. Pessimism is a self-fulfilling prophecy. If you don't enter the race, you can't finish. This reason for inaction is just another big bird with its head in the sand and ignores the enormous progress we've made solving environmental problems, such as reducing toxins in our air and water. The magnitude and complexity of biodiversity-related problems are certainly more daunting than others we have solved, but the solutions are at hand because we know and can control the leading cause of biodiversity loss—our behavior. We just need to figure out what behaviors sustain the desired biodiversity and foster the will to design economies that encourage us to act accordingly.

Substitutes: Faith that the ultimate resource—human creativity—can create substitutes for lost biodiversity must be balanced against the serious challenges of replacing nature's services. The pace of technological advance over the last few centuries provides ample reason for optimism, but the enormity and complexity of nature's services that sustain our lifestyles far exceed any problem we've previously conceived, much less solved. The extent of our ecological ignorance should humble the staunchest technophile; we know very little about the ecological functions each species performs and likely do not currently possess the technological capacity to replace these functions should they prove critical to human survival.

Even if we did possess the technological prowess to replace all of nature's services with our own engineering, the cost of doing so likely exceeds our ability to pay for it—the value of nature's services that humanity ungraciously accepts far exceeds the combined output of all the world's economies. Thus, we may possess neither the technological prowess nor the economic wealth to replace the life-sustaining services that biodiversity provides. Perhaps as important, biodiversity helps us define who we are as individuals, as communities, and as nations. It is a monument to our values, a moral mirror into which we look and see our priorities, and a source of identity. It may remind us of our connection to the Creator or our debt to our ancestors. It may inspire us to write poetry, live sustainably, or study science. The values and lessons of nature shape us as much as do the values of democracy, freedom, and justice for all.

Thus, we need reasoned discussions about which environmental qualities we can degrade and do without, which ones we can afford and expect to replace with technology, which ones are too risky to ignore and degrade, and which qualities are essential to our human psyche and cultural character. These questions are difficult to answer, but we must try. The answers are not found in universal implementation of the precautionary principle, but rather a deliberate consideration of when and where we want to exercise caution and when and where we will exchange risks caused by extinction for hopes of improving current and future living conditions for people. We should explicitly identify those environmental qualities we can live without and those for which no substitutes exist. Thus, the biodiversity we choose to sustain should reflect, at least in part, its contribution to essential life-support services, our comfort level with risk, our faith in our own creativity to replace it, and its contribution to our character and dignity.

Leave It in God's Hands: The charge that the Judeo-Christian land ethic encourages subjugating and remaking nature in our own image attracts considerable critique from environmentalists complaining that these attitudes sanction environmental degradation. Moreover, if God will provide until the Apocalypse arrives and Rapture begins, then doesn't Christianity encourage the use of resources as if they were infinite and hasten the Second Coming with environmental destruction, as some end-timers apparently advocate? Religious communities have responded vigorously to these critiques. Everyone from the Pope to "God's Greens" advocate caring for creation, equitably distributing environmental justice, tending and keeping the garden, and carefully stewarding God's Earth so that it can be passed on to future generations. Mainline religions now have written policies on everything from oil exploration in Alaska to sustainable agriculture in Texas. In addition to focusing on achieving spiritual life after Earth, the devout should also focus on being good shepherds of creation during life on Earth.

Lacking the Will to Act: The most plausible explanation for inaction seems to be that we just don't care about biodiversity. Or enough of us don't care. Or enough of us don't care enough. Deciding which extinctions to prevent is one thing; changing extinction-causing behaviors is another. Why should we act differently? Biodiversity loss just doesn't seem relevant when compared to pressing concerns about our health and security, stable democratic institutions, a thriving economy, and finding dignity and meaning in our lives. Biodiversity seems abstract, remote, and anti-human. It evokes images of birds and beetles living in vast old-growth forests where few people visit or live. That image does not resonate with suburban, commuting, Web-surfing Americans; worse, it creates a false dichotomy: biodiversity versus humanity. Biodiversity needs an image makeover. It needs to be understood as something local and relevant to middle-class America. It needs to be appreciated as the ecosystem services that clean water, generate oxygen, and moderate temperature. It needs to be linked to human health and community vitality. Biodiversity should be seen as an aesthetic force increasing property values and evoking awe and respected as a repository of history and as a connection to the Creator.

Solving any problem—from biodiversity loss to dysfunctional families—requires three things: the problem must be identified, a solution must be devised, and the problem solvers must have the willpower to see the problem and implement the solution.[2] We recognize biodiversity decline as a problem, and solutions have been proposed. What we seem to lack is the will to act.

Our tendency to polarize issues as supporting either humans or nature saps our will to act. Discussions typically get framed at the extremes of economics and ecology, ignoring many other natures. Biodiversity gets narrowly defined as an abstract ecological construct standing in opposition to human economy and culture. Discussions within this polarized decision space are limited to trading off human welfare with biodiversity—economy

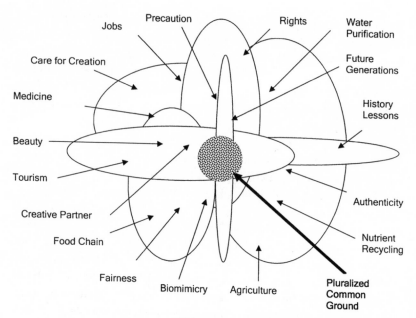

Jobs Precaution Rights Water
Purification

Care for Creation Future
Generations

Medicine History
Lessons

Beauty

Tourism

Creative Partner Authenticity

Food Chain Nutrient
Recycling

Fairness Biomimicry Agriculture Pluralized
Common
Ground

FIGURE 1.

versus environment or jobs versus owls. The polarized decision space makes
it seem that biodiversity thrives only when humanity leaves nature alone
and that humans thrive only at the expense of biodiversity.

By pluralizing our conception of biodiversity, we can promote different ways for people to understand and value it. By focusing negotiations
where these conceptions overlap, we can increase the number of motivated
stakeholders and move beyond the paralysis that polarization causes. For
example, the outline of many natures used to structure this book can help
us look for areas of common ground. Ecological concerns highlight the
interdependencies among species; impacts to one species are felt by many
others. Economic concerns remind us that many of these interdependencies provide services critical to human well-being and are expensive if not
impossible to replace. Health concerns illustrate the value in pharmaceu-
tical cures to human ailments contained within the chemical composition
and biological actions of other organisms. Bio-rights concerns highlight the
lives and rights of countless creatures protected and respected along with
diversity. And concerns about fairness and equity help us realize that biodiversity differentially affects the lives and interests of current and future
people.

Biodiversity also evokes profound aesthetic appreciation. Each species
has a multimillion-year story of survival to tell and provides a powerful
authentic connection to Earth's distant past. As well, biodiversity provides

a moral mirror reflecting our values: we might feel the self-loathing of obesity rather than the pride of fitness when, looking at our landscape, we see extinction after extinction caused by self-absorbed excess. In addition, biodiversity respects the creative potential of evolution. Each species represents accomplishment likely unparalleled by human inventiveness: we are better for having nature as a creative partner. Biodiversity also is linked with humanity. For millennia, especially after humans learned to control fire a million years ago, our extensive and pervasive modifications of the biosphere make us the stewards of biodiversity. Our self-awareness makes us responsible for how we carry out this stewardship.

Indeed, many natures do contain biodiversity and provide multiple reasons to care about it. Biodiversity should thrive on the higher ground where these many natures overlap.

PRAGMATIC PLURALISM

The environment is so complex and our abilities to understand it so imperfect that our appreciation of it will always be limited. The best we can do is assemble numerous partial glimpses—partial natures. Each nature appears to us through a different lens of human culture: science, art, religion, and so on. Each nature has people who see and appreciate it, and each nature defines a role for humans to play. Some natures thrive independently of humans; other natures thrive only through human understanding. Some natures define limits that humans exceed only at great risk; other natures are malleable and can be reconfigured through a respectful and deliberate partnership.

We must use civil discourse to choose among the many natures the ones we want to sustain, create, and extinguish. Pluralizing nature may help uncover shared interests, reveal potential political partnerships, and motivate community action. Rather than be polarized into apathy by fighting over whose nature is best, we can instead join forces to realize our shared interests. Many natures overlap. As is the case with biodiversity, physical conditions that satisfy the values and concerns of one nature likely satisfy the values and concerns of other natures.[3]

A restricted decision space causes frustration and distrust, limiting opportunities for civic discourse. People get excluded from discussions because they lack the language to describe the natures they value. Their natures and values get dismissed as unfounded opinions or personal tastes. Only "scientific facts" and "economic realities" merit serious consideration.[4] People excluded from discussions understandably search for ways to circumvent the deliberative, democratic process with political protest, obstructionist lawsuits, and monkey wrenching. Rather than excluding people, we should be including people. The more people care about nature, the more we are likely to mobilize the political support needed to rein in unsustainable behaviors and create thriving communities. Pluralizing nature legitimizes many voices by opening the decision space.

Even though it may seem like environmental heresy to say so, I believe human creativity can improve, refine, and enhance many natures. I value natures that help humans feel positive about our roles in the biosphere and thus inspire the will to act as stewards, partners, and lovers of nature. I think humans can be more than responsible citizens of the biotic community; we can be prudent innovators, inspired visionaries, and loving partners in the odyssey of evolution. I can feel this way without losing respect and appreciation for the creativity and complexity of natures that are autonomous, wild, and rightful.

Of course, we always must be wary of hubris. We must admit that we can destroy as well as create many natures. We must be cautious of our own creativity. We must acknowledge that our technological prowess and growing population amplify our ability to eliminate natures, while our capacities to envision, value, and sustain natures remain limited. Many natures exist, many more are possible, and many of these natures can have a nurturing, creative role for humans. Thriving and sustainable communities exist on the higher ground where these natures overlap.

Notes

CHAPTER ONE

1. Shellenberger and Nordhaus (2004). Adam Werback, past president of the Sierra Club, furthered this critique by asking: "Is environmentalism dead?" in a controversial December 2004 speech (http://www.grist.org/news/maindish/2005/01/13/werbach-reprint/index.html). See also Botkin (2001), Dunlap (2004), and Meine (2004) for a more historical view of the critique.

2. Senecah (1996) reviews a particularly polarizing planning negotiation for the Adirondacks where the name-calling was especially creative.

3. Thoreau is discussed in more detail in chapter 13; Leopold is discussed in chapter 4.

4. Leopold (1967). "Ecological Conscience," in Flader and Callicott (1991, 346).

5. "Community" is used here in a Leopoldian sense and as further developed by Norton (1990, 2005) and Flader (2004).

6. The is-ought debate and naturalistic fallacy are discussed throughout the book. For discussion about it and our postmodern search for meaning in nature, see Evernden (1992).

7. Dukes et al. (2000); Dryzek (2000); Lewicki, Gray, and Elliot (2003); Schon and Rein (1994).

8. "Flabby," "fortress," and "engaged pluralism" are terms used by Bernstein (1988). Engaged pluralism goes by many names and comes in various forms: civic environmentalism (Shutkin 2000), republican environmentalism (Posner 1996), public ecology (Robertson and Hull 2003b), ecological democracy (Ferry 1995; Gundersen 1995), constructing sustainability (Norton 2005), and discursive democracy (Dryzek 2000).

9. Opie (1998).

10. Curry (2003) reviews critiques suggesting that constructionism and postmodernism weaken environmental protection. Snyder (1998) argues that some deconstruction is nothing more than "the high end of the 'wise use' movement," "at the right time to bolster global developers" and thus hasten environmental degradation. However, in the same paper, he argues that "deconstruction, done with a compassionate heart and the intention of gaining wisdom," allows one "to examine one's own seeing so as to see

the one who sees, and thus make seeing more accurate." I believe pluralizing nature helps us see our partnership with nature.

CHAPTER TWO

1. Retracing and reinterpreting de Soto's journey remains an active pursuit of historians. His ambitions for Lake Michigan remain debated. Milanich and Hudson (1993) retrace de Soto's expedition using journals to document the presence of villages, roads, bridges, agriculture, and other signs of civilization.

2. Krech (1999) estimates that most of the native North American population lived in the Southwest, the Northeast, California, and the Southeast. Other estimates for North America range between 2 and 20 million.

3. Bonnicksen (2000), Denevan (1992), Diamond (1999), MacLeish (1994), Flannery (2001), Krech (1999), and Simmons (1989) provide a general overview of the extent of Native American civilization. Agriculture was widespread, spanning the whole continent. That is not to say that all Native Americans formed agricultural communities or that agricultural fields stretched uniformly from coast to coast. Many communities remained hunter-gathers, and vast acreages remained unfarmed if not unburned and unhunted.

4. Mink (1992).

5. Smith (1996) describes the accomplishments and demise of the Aztec empire. Diamond (1999) provides readable conjecture about why the Aztec and other empires rose and fell.

6. Phrasing and advocacy of the "noble savage" is often attributed to Jean-Jacques Rousseau (1712–1778). Pinker (2002) reviews this and contrasting visions of human nature. The Grinnell paper and romantic myths of the noble savage are discussed by Smith (2000). Also see White (1984).

7. The history of the Chief Seattle quote is discussed in Knudtson and Suzuki (1992, xv–xvii). Apparently there exist five or six versions of the speech, none of them verbatim. Each version reflects more and more Eurocentric ideas. The most quoted version is supposedly an entirely fictionalized script from a documentary film on pollution.

8. Krech (1999) reviews evidence for and against ecological Indians. See also Redman (1999) and White (1984), who focus on cultural norms promoting both environmental protection and degradation. Cronon (1983) argues that Native Americans did not consciously limit their environmentally destructive behaviors, but Martin (1987) contends that by imbuing spirituality in animals, Indians minimized takings beyond that expected based on a purely economic model.

9. LeBlanc (1999) argues that warfare was common and likely motivated by the struggle over scarce resources. He pointedly attacks the myth of idyllic aboriginals living off the bounty of nature. He concludes from archaeological evidence that the unsustainable land-use practices of aboriginals around the world motivated aggressive behavior and territorial disputes that resulted in warfare being common. Diamond (2005) extends this argument with examples of past and current cultures from around the world.

10. Van Tilburg (1994) maintains that the cause for social collapse was much more complicated than environmental degradation but acknowledges and documents the loss of soil fertility and deforestation that no doubt contributed to the dramatic cultural chaos. Dutch explorers discovered Easter Island in 1722. They reportedly shot many tribal leaders at first sight and caused great disruption to the isolated culture. See also Diamond (2005).

11. Gomez-Pompa and Kaus (1992) argue that primitive technologies differ from modern technologies in their potential environmental impacts because they evolved and were tested in local conditions. Local technologies that failed to create a sustainable land practice were culled by natural selection. Modern technologies, in contrast, have greater potential for harm because they are not subject to such tests. Diamond (1999), however, contends that primitive technologies were not all that local but were widely shared. He suggests that the ease with which technology was shared in Europe and Asia explains why European settlers enjoyed a technological edge over native cultures in America, Australia, and Africa, where transport of technology was blocked by mountains, deserts, and other physical barriers.

12. Vitousek et al. (1997); Steinberg (2002); MacLeish (1994).

13. McKibben (1989).

CHAPTER THREE

1. The description of how nature has changed over the last few thousand years was extracted from comprehensive treatments by Flannery (2001), Bonnicksen (2000), MacLeish (1994), and Diamond (1999).

2. Gobster and Hull (2000).

3. For accessible readings sympathetic to evolution as an explanation of life on Earth, see Dawkins (1996), Goodenough (1998), Gould (1989), Futuyma (1998), and Mayr (1988). Theoretical mechanisms of evolution are hotly debated. For example, punctualists argue that some mutations offer no advantage but get carried along until they accumulate with other mutations to a point where they offer an advantage or detriment. Gradualists, in contrast, argue that each mutation must offer some adaptive advantage if it is to be carried on to the next generation. There is also debate over whether the pressures of selection are exerted only on organisms or also on groups, thus raising the questions of whether groups can evolve because they share traits that facilitate survival and reproduction. Such a debate is particularly relevant to questions of whether humans evolved traits of community and altruism that benefited the survival of the group. The references cited provide details about these debates. These details are glossed over here because they are not essential to the chapter's major point that one powerful way of understanding nature is seeing it as the result of random, unintentional changes that grant reproductive and competitive advantages.

4. The movie of life analogy is from Gould (1989), whose interpretation of evolution emphasizes randomness and contingency. See the next note for an alternative emphasis.

5. The potential for life is constrained by the realities of physiology, the demands of the environment, and the potential present in the most recently successful genetic systems. New life-forms are at most modest mutations from parents and must prosper in conditions that tested and produced life as we know it. The eye of a squid and the eye of a human perform remarkably similar functions but do so using slightly different mechanisms derived following different evolutionary paths. The convergence on function is used to suggest that certain abilities, such as vision, are likely to be present in many life-forms on Earth. Morris (1998) responds to Gould's heavy emphasis on randomness: "Convergence demonstrates that the possible types of organisms are not only limited, but may in fact be severely constrained. The underlying reason for convergence seems to be that all organisms are under constant scrutiny of natural selection and are also subject to the constraints of the physical and chemical factors that severely limit the action of all inhabitants of the biosphere" (202). Wright (2000) argues that evolution "progresses" toward higher and higher levels of complexity and information processing.

6. The "voyagers" and "odyssey" phrasing comes from Leopold (1967, 109).

7. Gould (1989); Farber (2000).

8. Farber (2000); Norton (2003, 58); Mayr (1988); Sagoff (1988b).

9. Farber (2000)

10. See http://www.ohiou.edu/phylocode.

11. Diamond (1999); Smith (1998); Pollan (2001).

12. The version of social Darwinism presented here for illustrative purposes emphasizes the competitive qualities of natural selection. It might better be called social Spenceri-
anism, because it represents the particular interpretations of Darwin by Herbert Spencer.
Spencer supposedly coined the phrase "survival of the fittest." He argued that competition
necessarily eliminated unfit individuals from the population and that social programs allowing inferior and unworthy individuals to reproduce do damage to society because they
maintain inferior traits in the population that natural selection would otherwise eliminate.
Spencer had a radical individualism, anti-government, laissez-faire political philosophy
long before he adopted (and adapted) Darwin's theory of evolution as justification for
these policies. Loaded terms such as "unworthy" are taken from Spencer's writings as
presented in Hawkins (1997). For a historical review of the many efforts to construct ethical systems and moral principles from evolutionary theory, see Farber (1994) and Midgley (1985, 1995). Bannister (1979), Collins (1959), and Hawkins (1997) provide excellent
accounts of the various interpretations and critiques of social Darwinism. Degler (1991)
traces how evolutionary thought has shaped and changed biological and social sciences
in America.

13. Hawkins (1997, 242).

14. Pinker (2002).

CHAPTER FOUR

1. I took some literary license and modified Leopold's account of X's journey in the
essay titled "Odyssey" (1967, 104-7).

2. Beliefs that nature knows best and seeks a balance are still popular lay understandings of nature: see Fitzsimmons (1999), Kempton, Boster, and Hartley (1995), and
Hull et al. (2002, 2003). Botkin (1990), Egerton (1973), and Shrader-Frechette and
McCoy (1995) review and critique the notion of balance in ecological science. Important efforts to redefine ecological health and integrity without appealing to balance are
being conducted by resilience organization, Gundersen and Holling (2002).

3. There are numerous excellent treatments of the models or paradigms by which
nature has been interpreted within ecological science, including Bocking (1997), Botkin
(1990), Merchant (1989), McIntosh (1985), Sagoff (1988b), Worster (1994), and Wu and
Loucks (1995). The three models presented here are a simplification of a much richer
and more complicated history of debate over the correct way to understand ecological
interactions.

4. Norton (1993) differentiates between strong and weak holism. Strong holism implies that the whole's essence is spiritual and perhaps even willful. Weak holism suggests
that the whole is simply more than the sum of its parts because feedback loops, dissi-
pative structures, and other properties of systems emerge from interactions among the
parts. Definitions of "ecosystem" are hotly debated in part because of different opinions
about holism, mechanical, and disturbance-equilibrium models of nature. Some place the
ecosystems concept at the center of ecology, others at the nonscientific fringe (Bocking
1994; Worster 1994).

5. Clements's quotes are taken from Worster (1994, 211). While these words suggest literal holism, some have argued that Clements really was using organismic analogies as a heuristic.

6. Gundersen and Holling (2002); Norton (2003).

7. Levins and Lewontin (1985, 211).

8. McQuillain (1993) characterizes and contrasts organismic and mechanistic models of forestry. The maximum mean annual growth refers to the increment of merchantable wood a tree accumulates as it grows taller and thicker. As the rate of growth starts to slow, the time for harvest nears because younger trees put on wood (and hence profit) more quickly. Trees treated in such a way are "the floral equivalent of faunal veal" (192). See Langston (1995) and Chase (1995) for detailed histories of forestry that contrast organismic and mechanistic models.

9. The history of the Blue Mountains is examined in Langston (1995). See Gundersen, Holling, and Light (1995) for analysis of numerous case studies of how ecological disturbance and hierarchical theory provide appropriate ways of thinking about natural resource management. Gundersen and Holling (2002) integrate hierarchical ecology, economics, and sociology to form a theory of panarchy that strives toward resilience rather than stability and control.

10. Bocking (1994); Callicot, Crowder, and Mumford (1999).

11. "Biological Integrity" is mandated in the 1972 Clean Water Act. The quote defining integrity is from Karr and Dudley (1981), emphasis added. Important discussion and debate of integrity can be found in Angermeier and Karr (1994), Karr (2002), and Westra and Lemons (1995).

12. Woods and Moriarty (2001); Sagoff (1999); Gobster (2005); Bright (1998).

13. Meine (1988) provides a respected biography of Leopold.

14. Quotes about cultural harvest, love, and slavery come from Leopold (1967, ix, 224, 225). Arguments about promoting perception are on pp. 172–74 and discussed below. Also see Meine (1988, esp. p. 350). The penultimate statement (a thing is right) occurs at the end of Leopold's "Land Ethic" essay (1967).

15. The quotes from Leopold used in the previous paragraphs occur at the very end of *A Sand County Almanac* (1967) in the "Land Ethic" essay. His definition of community, ethics, and obligation occur at the beginning of that essay. The explicit quotes about humans being fellow members, part of a biotic team, and respecting the community as such occur on pp. 202–4. Elsewhere Leopold critiques economically motivated science that "can invent more and more tools, which might be capable of squeezing a living even out of a ruined countryside, yet who wants to be a cell in that kind of a body politic? I for one do not" (quoted in Flader and Callicott 1991, 217; "Taken from Land Pathology"). Callicott explains and defends Leopold in numerous essays in chapters 4, 5, and 6 of his 1989 compendium (*In Defense of the Land Ethic*). He traces Leopold's logic back to Darwin and Hume, who argue ethical concern begins with inherited concern for protecting offspring and others that can help protect offspring. Norton (1988, 2003, 2005), in contrast, traces Leopold's logic to the pragmatist reasoning of Arthur Hadley, the president of Yale during Leopold's studies. Leopold likely read Hadley's philosophical treatise that contained a summary statement strikingly similar to Leopold's penultimate statement: "The criterion which shows whether a thing is right or wrong is its permanence. Survival is not merely the characteristic of right; it is the test of right." Although there exist competing explanations for Leopold's logic, I argue that bridging the human-nature dichotomy and focusing our attention on the large context are two of Leopold's greatest accomplishments. I find Norton's interpretation of Leopold more constructive

because it proposes a method for identifying (negotiating) the larger contextual units and the qualities in them that should be sustained. Leopold biographer and scholar Flader (2004) also traces Leopold's emphasis on community values and citizen responsibility in contrast to individualism, capitalism, and even professionalism and property rights.

16. "Conservation Ethic" was published in the *Journal of Forestry* and in Leopold (1993).

17. Additional challenges and charges of fascism are discussed later in the comparison of biocentrism and ecocentrism (chapter 11: "Rightful Nature").

18. Quotes are from Leopold's essay "Conservation Aesthetic" in *A Sand County Almanac* (1967, 174).

19. Leopold (1967, 96–97); he talks about loving the land in several places and directly on p. 223.

20. Sears (1964); Shepard and McKinley (1969). See also Hayward (1994).

21. McIntosh (1985) and Worster (1994) review the history of ecology. In Britain the Nature Conservancy was established by government funding as an independent institution to support ecological research (Bocking 1997). It was a nongovernmental organization in the United States.

22. Murdoch and Connell (1973, 169).

23. Various articles in *Science* 287 (February 2000): 1188–95. See also Robertson and Hull (2003b) and *Bioscience* 51, no. 6 (2001).

24. For an overview, see Harvey (1993), Porritt (1985), or Dryzek (1997). For deep ecology, see Ness (1973). For bioregionalism, see Sale (1985). For sustainability, see Dryzek (1997) and Peterson (1997). For social ecology, see Bookchin (1989).

25. Fitzsimmons (1999).

26. Chase (1987, 1995), Huber (1999), Kaufman (1994), and Lewis (1992), in that order.

CHAPTER FIVE

1. Lily-pond analogy taken from Brown (1978).

2. Mumford (1961); Gottlieb (1993) traces the roots of environmental concern to urban and social conditions.

3. Quotes from Malthus's essay taken from Worster (1994, 150). Linner (2003) reviews the history and implications of Malthusian thinking.

4. Marsh (1885, 9). See Lowenthal (2000) for a detailed biography of Marsh.

5. Price (1999) reviews the passenger pigeon extinction. Barrow (2002) reviews the history of "extinction" as a conservation idea.

6. Worster (1979).

7. White (1995); Scarce (2000).

8. The metaphors come from popular books and articles: explosion (Ehrlich 1968), spaceship and cowboy (Boulding 1973), and lifeboat (Hardin 1974).

9. Meadows, et al. (1972).

10. See Bast, Hill, and Rue (1994) and Lewis (1992) for technological success stories and Sedjo (1995) for statistics about natural areas and forests.

11. Pro-technology arguments can be found in Easterbrook (1995), Huber (1999), Lewis (1992), and Simon (1981).

12. Barnett and Morse (1963); Simon (1981); Lomborg (2001).

13. Or overfishing the oceans, polluting the air, or affecting other domains owned by no one and accessible to all. See Hardin (1968).

14. The Bruntland Report (World Commission on Environment and Development 1987, 43). Sustainability is discussed throughout these chapters, especially the next chapters on economics and equity.

15. Cross (1997); Myers (2002); Posner (2004).

16. Allen, Tainter, and Hoeskstra (2003, esp. chap. 2) summarize studies that document how the gains from technological advances in agriculture, medicine, and science generally have steadily declined while costs have increased.

17. The hard-green position is well represented by Huber (1999); the soft-green position is well represented by McDonough and Braungart (2002) and Bookchin (1989).

18. Benyus (1998) describes bio-mimicry. *The Machine in the Garden* is the title of a book by Marx (1964) that examines technology and the pastoral ideal in American culture.

19. McDonough and Braungart (2002, 15-16).

20. Bookchin (1989); Kirk (2001).

21. Worster (1994); Hays (1959, 1987); Dryzek (1997).

22. Hays (1987) and Worster (1994) document the changing language of environmental science. Sagoff (1988b) explains the scientific limits of describing nature. See also Hull et al. (2002).

23. Dutney (2001).

24. Eden (1998); Lemons (1996); Wynne (1992); Yearley (2000).

25. Ehrenfeld (1981) provides a classic treatment of this topic. Posner (2004) updates our dilemma with detailed discussion of risk.

CHAPTER SIX

1. Marx (1964), Smith (1950), and Steinberg (2002) provide excellent reviews of this history and numerous examples of how economics shapes our use and understanding of nature.

2. The army metaphor comes from William Gilpin, a strong advocate of Manifest Destiny in the mid-1800s: "Each family was a platoon, making a farm 'on the outer edge of the settlements.' 'As individuals fall out from the front rank, or fix themselves permanently, others rush from behind, press to the front, and assail the wilderness in their turn.' Thus, a 'tidal wave of population' defeats 'the glebe, the savages, the wild beasts of the wilderness, scaling mountains and debouching down upon the seaboard'" (Gilpin quoted in Stoll 1997, 121).

3. The "scientific" observations of Alexander von Humboldt, a German geographer, further justified this vision of Manifest Destiny. His theory predicted that world empires migrated westward, starting centuries ago with China and India and migrating west through Persia, Greece, Rome, Spain, and on to Britain. The sinuous line connecting the geographic pattern pointed next to North America. (Smith 1950, chap. 3).

4. So argues Smith (1950, chap. 14).

5. Smith (1950, 163), quoting from *The Writings of Abraham Lincoln*, 8 vols., ed. A. B. Lapsley (New York: Putnam's Sons, 1905-6), 6:194.

6. Teddy Roosevelt (1910, 262-63, 269-70). See Steinberg (2002, 81) for context. Buffalo eradication was also justified as an act of war to remove the food source of Native Americans.

7. Pollan (2002a, 2002b) provides a detailed and readable description of industrial meat production. The feeding of animal parts to cattle has been questioned and discontinued in some areas due to concerns about mad cow disease.

8. Klinkenborg (2001) traced the flow of cow parts in his research examining impact of mad cow disease.

9. American Sheep Industry Association, http://www.sheepusa.org (accessed Jan 2003).

10. The quest to define sustainability and the logic of sustainable development reviewed in Dryzek (1997), Hajer (1995), Newton and Freyfogle (2005), and Peter (1997).

11. Capitalism is reviewed and critiqued in Heilbroner and Galbraith (1990), L (1999), and Weber (1930).

12. Grossman and Krueger (1995) support the Environmental Kuznets Curve (EK Arrow et al. (1995) and Rothman (1998) critique the EKC.

13. The Bruntland Report (World Commission on Environment and Developm 1987, 43).

14. See Norton (1998a, 2005). Issues of sustainability are also discussed later chapter 8: "Fair Nature."

15. Hawken (1993).

16. The story of watering New York is told in Daily and Ellison (2002) and V (2000). Sagoff (2004, chap. 6) disputes parts of the story, suggesting that much needed doing than environmental services advocates purport and that New York City transferred much less wealth than promised.

17. The Costanza et al. (1997) estimate of $18 to $61 trillion is admittedly tenta and not without serious critique (Sagoff 1997). Balmford et al. (2002) also argue tha makes economic sense to preserve intact ecological services (with a benefit cost rati 100 to 1).

18. Allen (1991); Luke (1997); http://www.bio2.com.

19. McPherson (2003).

20. Daily and Ellison (2002, 232). The green-thumb quote is attributed to E. Wilson by Daily and Ellison (2002, 16).

21. Laroche, Bergeron, and Barbaro-Forleo (2001), Frankel (1998), Kaufman (199 Menon et al. (1999), Ottman (1988), and Prothero and Fitchett (2000) provide overview of the green market. See also GreenBiz.com, an organization devoted to busi capturing this market.

22. The EPA's Environmentally Preferable Purchasing program is reviewed http://www.epa.gov/opptintr/epp/index.htm; The UN's Sustainable Consumption gram is reviewed at http://www.unep.fr/en/branches/pc.htm. The quote was accesse June 2003.

23. McDonough and Braungart (2002). Cradle-to-cradle philosophy is discusse more detail in chapter 5.

24. The Consumers Union guide to environmental labels: http://www.eco-lab org/home.cfm.

25. Tierney (1996) takes some cracks at many recycling myths in a *New York Ti* article. The Styrofoam-ceramic debate remains contested but raises good questions ab the difficulty involved in making the "right" choice.

26. Frankel (1998), Freeman, Pierce, and Dodd (2000), Hoffman (2000), Schmidheiny and Zorraquin (1996) review motivations for business to green. "Env preneurially" was coined by Menon et al. (1999), who also review motivations for gr business.

27. Hays (1987).

28. The so-called Superfund sites; see Schmidheiny and Zorraquin (1996, chap.

29. Home Depot: www.homedepot.com/HDUS/EN_US/corporate/corp_respon/wood_purchasing_policy.shtml (accessed March 2004). Also see Hornblower (1998). The International Trade Forum describes a similar story for a large home improvement chain in Britain.

30. Design for the Environment (http://www.epa.gov/dfe) and Pollution Prevention Pays (http://www.epa.gov/oppt/p2home/index.htm) are Environmental Protection Agency programs that illustrate how businesses can develop new products or redesign old ones in ways that reduce health risks and pollution costs.

31. Romm (1994).

32. Hoffman (2000, chap. 4); http://www.ceres.org.

33. The Declaration of the Business Council for Sustainable Development is reprinted in Schmidheiny et al. (1992, xi).

34. Frankel (1998); Freeman, Pierce, and Dodd (2000); Hawken (1993); Schmidheiny and Zorraquin (1996).

35. Hajer (1995).

36. Sagoff (1981, 1988a, 2004).

CHAPTER SEVEN

1. Carson (1962, 22).

2. Gottlieb (1993; esp. chap. 2); Hays (1987); Sale (1993).

3. Freeze (2000); Philp (1995).

4. Gibbs (1998); Philp (1995, 80–83).

5. EPA website, http://www.epa.gov/ebtpages/humanhealth.html (accessed February 2003).

6. Lilliston and Cummins (1998).

7. Graham and Wiener (1995, 12–13); Gray and Graham (1995, esp. 186–87). See also the discussion on labeling and green marketing in chapter 6.

8. U.S. Consumer Product Safety Commission: Fact Sheet on Chromated Copper Arsenate (CCA)-Treated Wood Used in Playground Equipment, http://www.cpsc.gov/phth/ccafact.html (accessed February 2003).

9. Wilson (1992).

10. Driver, Brown, and Peterson (1991); Ulrich (1984, 1993).

11. Quotes are by former Harvard president Charles Eliot in 1914 and architect Warren Manning in 1901 as cited in Schmitt (1990, 73, 71).

12. Olmsted also believed urban parks could enhance the democratic ideals of community. Parks should be places where people of all social ranks gathered and mixed, creating a sense of community and shared responsibility that would nourish democracy.

13. Statistics about early park acreages are from Schmitt (1990, 70–74). Current standards are set by each state during the planning process for SCORP (State Comprehensive Outdoor Recreation Plan). Each state completes a SCORP to remain eligible for federal funds.

14. Sellars (1997, chap. 2).

15. Driver, Brown, and Peterson (1991); Parsons et al. (1998).

16. Beck (1992); Dryzek (1997).

17. Freeze (2000) and Philp (1995, chap. 2) discuss risks from toxins. Beck (1992) and Dryzek (1997) discuss living in a risk-based society. Hays (1987, chaps. 6, 11) discusses the toxic burden and knowing environmental quality through the lens of health; EPA website: http://www.epa.gov/epaoswer/osw/tsd.htm.

18. Over time these animal tests are compiled with field data of accidental exposures and medical reports of human sensitivities to refine the safe exposure level (Gray and Graham 1995, 175–76).

CHAPTER EIGHT

1. Bookchin (1989); Bullard (2000); Gottlieb (1993); Shutkin (2000).

2. The memo was from Lawrence Summers and was reprinted in the *Guardian*, February 14, 1992, p. 29.

3. J. P. Ross, CorpWatch, article 2588 (2002), http://www.corpwatch.org/article. php?id=2588 (accessed March 2004).

4. These and other examples are cataloged by the Center for Investigative Reporting (Moyers 1990). For more provocative treatment, see Rachel's Hazardous Waste News #126, April 25, 1989, http://www.ejnet.org/rachel/rhwn126.htm. For a technical discussion, see Freeze (2000).

5. Yearley (1996).

6. The history is excerpted from the preface of Cole and Foster (2001).

7. The quote was made by Horace Kephart, often called the father of the park. The quote and most of this history of the Great Smoky Mountains National Park is taken from Weaver (1996, 163), who, in turn, quotes from Michael Frome's 1966 history of the region.

8. The negative stereotypes are discussed in Weaver (1996). Robertson and Hull (2003a) provide a broader discussion of alternative understandings of the southern Appalachian region.

9. Volume 8 of the *Duke Environmental Law and Policy Forum* reviews many issues of willingness to take risk affecting environmental policy, in particular see Cross (1997).

10. Norton (1998a); Allen, Tainter, and Hoekstra (2003).

11. Passmore (1980) argues against obligations to the future, while Norton (2005) argues for obligations. Beatley (1994) reviews several of these arguments.

12. See Gottlieb (1993) for a discussion of equity concerns in the early environmental movement, which started with urban concerns about oppressive and dangerous living and working conditions.

CHAPTER NINE

1. Of the one thousand people telephoned in a Gallup poll on March 5, 2001, 45 percent believe that God created human beings pretty much in their present form within the last ten thousand years or so. Thirty-seven percent believe that human beings developed over millions of years from less advanced forms of life but that God guided this process. Twelve percent believe that God had no part in this process. The Gallup poll was conducted for the Roper Center for Public Opinion Research (Storrs, CT) and is distributed by Polling the Nation, http://poll.orspub.com/poll/lpext.dll?f=templates&fn=main-h.htm (accessed September 12, 2005) and copyrighted by the Los Angeles Times Syndicate.

2. Most of the chapter restricts its focus to Protestantism. Stoll (1997) argues that the contribution of Protestantism to American environmental thinking is profound but largely goes unnoticed. He argues that rationality, individualism, anti-authoritarianism, moral activism, feminism, capitalism, and environmentalism spring from and are nurtured by Protestant thought. Of the Ten Commandments, some scholars of environmental

theology suggest that the commandment to honor the Sabbath has the most environmental implications because it commands that creation—land and humans—rest and worship every seven days. Excellent sources for environmental theology include Callicott (1994), Cherry (1980), Clark (1994), Fowler (1995), Glacken (1967), Oelschlaeger (1994), Passmore (1980), Peterson (2001), and Turner (1989).

3. White (1967) offers one of the most quoted critiques of Christian environmental philosophy, arguing that it promotes a dualism that encourages human domination over nature, which presumably contributes to degradation of environmental quality. Other accessible and important analyses of dominion include Peterson (2001), Stoll (1997), and Passmore (1980).

4. Italics added. The second telling begins in the second half of Genesis 2:4 and is discussed in the next section.

5. Black (1970); Passmore (1980).

6. Marx (1992).

7. A quote attributed to Smohalla, a Lakota spiritual leader under pressure to adopt a Euro-American lifestyle; quoted in Callicott (1994, 121).

8. Thomas (1983) reviews early British views of nature and the rise of religious commentary on the subject, especially in the 1700s. The discussion of early worship of God in nature is discussed on pages 215–16 and elsewhere in chapters 4 and 5 of Thomas. See also Scully (2002) for a more popular treatment.

9. These and other interpretations can be found in Bailey (1917), Black (1970), Callicott (1999, chap. 10), Passmore (1980), Swartz (2002), Oelschlaeger (1994), Peterson (2001), and Scully (2002).

10. Berry, a powerful and prolific writer, has many publications developing a Christian environmental ethic. His essay in 1993 provides a focused critique of dualism. The quote is on p. 49. Callicott (1994) goes even further to argue that the original sin was/is dualist, anthropocentric thinking (17–20).

11. Points 1 and 7 of "The Ecological Crisis: A Common Responsibility" by Pope John Paul II. Message of His Holiness for the celebration of the World Day of Peace, January 1, 1990, http://www.ewtn.com/library/PAPALDOC/JP900101.htm (accessed September 12, 2005).

12. Botkin (1990); Worster (1994). Jefferson changed his beliefs about extinction later in his life. See Browers (1999); Miller (1988).

13. Calvin (1950, 52).

14. Fowler (1995); Peterson (2001); Stoll (1997).

15. Thomas (1983). See also Lutts (1990).

16. Stoll (1997, 80), quoting Edwards from his autobiography.

17. Muir quotes taken from Stoll (1997, chap. 7). Muir's pleas inspired enormous support, but construction on the dam began in 1914, just before his death. See Nash (1982) for additional quotes and more detail about the Hetch Hetchy watershed controversy.

18. Browers (1999); Miller (1988); Stoll (1997, 90–98).

19. The quotes come from Schmitt (1990, 141, 144); see also Lutts (1990).

20. Seton (1907, 24). Lutts (1990) summarizes other examples from this literature. Seton was lumped in with other "Nature Fakers" by Teddy Roosevelt and John Burroughs, who soundly criticized the methods and purposes of those writers.

21. Seton (1907, 31).

22. Ibid., 25–26.

23. Glacken (1967, esp. 293); Worster (1994, esp. 30).

24. Gifford Pinchot offers a Christian justification for his life of public service and his advocacy of environmental conservation (i.e., the greatest good for the greatest number for the longest time). Stoll (1997) discusses Pinchot's Puritan and Calvinist motivations in chapter 7. Pinchot (1910) wrote about "the Kingdom of God on earth" and similar religious metaphors in his book *The Fight for Conservation*. The quote is from pp. 95–96.

25. Cotton Mather quoted in Huth (1957, 7). See Stoll (1997) for a review of the Protestant work ethic and its effect on the American landscape.

26. Weber (1930) writes about Protestantism generally, not just Calvin.

27. Both quotes are from http://www.raptureready.com/index.php (accessed January 10, 2005). The Armageddon Clock is based on evangelical television pastor John Hagee, who authored bestsellers such as *From Daniel to Doomsday: The Countdown Has Begun* (1999).

28. Polling done for *Newsweek*, May 13, 2004. Quotes from http://www.leftbehind.com (accessed January 11, 2005).

29. Aukerman (1993, 27).

30. Boyer (1992).

31. Boyer (1992) and Rossing (2004). Scherer (2004) presents a more sensational account.

32. Lee (1995), reviewed and critiqued by Taylor (1999).

33. Rossing (2004) deliberately and methodically deconstructs the Left Behind and other end-time stories from a Christian perspective using Scripture as well as established theology.

34. Presbyterian Church (1990, 1996).

35. Some of the history is taken from the history page of the National Religious Partnership for the Environment (http://www.nrpe.org/history.html). "Earth and Faith" was published by the United Nations Environment Program, New York. *TNC Virginia Chapter News* (Spring 1999): 14. The *Outside* magazine article was written by Barcott (2001). Collaboration of the NCCC and the Sierra Club is described on http://www.ncccusa.org/news/02news1.html (accessed October 6, 2003). Worldwatch Paper 164 was written by Gardner (2002).

36. Oelschlaeger (1994) and Callicott (1994) are important works of this type. Peterson (2001) reviews and extends much of this work.

CHAPTER TEN

1. Thomas (1983) contends that the explicit acceptance of the view that the world does not exist just for humans should be regarded as one of the greatest revolutions in modern Western thought.

2. Vitousek et al. (1997) review the extent of human domination of Earth's ecosystems (see chap. 2).

3. See Budiansky (1998), Dawkins (1998), Donald (1991), Kitcher (1985), Eibl-Eibesfeldt (1989), Midgley (1995), and Preston (2002) for discussions of these and other continua along which humans and other animals are connected and differentiated. Definitions of terms such as *consciousness* are important parts of most debates about human nature. Terms can be defined inclusively to blur the distinction between humans and nature or defined exclusively to sharpen the distinction.

4. Arguments sympathetic to animal awareness of feelings can be found in DeGrazia (1996), Midgley (1995), Rollin (1989), and Scully (2002).

5. Whether animals possess and use consciousness depends on how consciousness is defined. Not surprisingly, many definitions of consciousness have been developed, some of which seem to intentionally advance the specialness of human nature by defining consciousness as the quality that distinguishes humans from other animals. Dawkins (1998) differentiates thinking, which evidences consciousness, from cleverness, which combines a good memory with the ability to apply simple rules of thumb through trial and error to succeed at tasks that offer rewards or avoid punishment. She defines "thinking" as requiring two conditions: the thinker should have an internal representation of the world in its head, and the thinker must be able to respond to changes of that representation (105). For arguments favoring consciousness in animals, see DeGrazia (1996) and Midgley (1994, 1995). For arguments against animal consciousness, see Budiansky (1998), Dennett (1987), and Heyes (1998). For a general treatment of the evolution of the mind in humans and animals, see Donald (1991) and Shettleworth (1998).

6. Defining and proving that "meanings" exist is exceptionally difficult, even in humans. Debates between behavioral and cognitive psychologists continue.

7. Prairie dog vocalizations differ for different predators: humans, hawks, coyotes (Slobodchikoff 2002). See Shettleworth (1998) for a discussion of alternative explanations of what seems to be intentional communication.

8. Behaviorists would argue that the chimp merely learned several new stimuli, rather than the generalizability of "more" (i.e., "more tickle" just became another stimulus for tickle). Even advocates of animal language are cautious of the bias in many experiments that credit language when it was not deserved. Terrace (1979); Terrace et al. (1979); Wallman (1992).

9. Donald (1991).

10. Pinker (2002); Tomasello (1999).

11. Examples of deception are reviewed by Waal (1986). The critique that the data is anecdotal and hence suspect to anthropocentric bias is reviewed by Whiten and Byrne (1988).

12. See chapter 9, "Spiritual Nature," for references and more discussion; see also Glacken (1967), Lovejoy (1960), and Stoll (1997).

13. The emphasis on rationality is consistent with Kant and other influential philosophers. Kurtz (1994) advocates humanism as a worldview. Evernden (1992, 31-35) describes the implications of humanism for understanding nature. B. F. Skinner depicts a humanist utopia in *Walden Two* (1948).

14. Curry (2003). Also see Peterson (2001), who argues that extreme constructionism is at least as dualist as Christianity.

15. Terrace (1979); Terrace et al. (1979); Budiansky (1998); Wallman (1992). See also Shettleworth (1998).

16. Premack and Premack (2003); Tomasello (1999).

17. Donald (1991) compares human and animal minds in a fascinating exploration of the origins of the human mind. He concludes his discussion about language by acknowledging that animals possess some rudimentary language abilities but adds that the "human linguistic superiority appears safe for the moment" (129).

18. Pinker (2002) and Ridley (2003) provide provocative and readable reviews of the recent history and current state of the nature versus nurture debate. Curti (1968) reviews competing definitions of human nature found throughout history in American philosophical and social traditions.

19. Eibl-Eibesfeldt (1989) reviews human ethology. Wilson (1978) reviews sociobiology. Degler (1991) documents how biological explanations of the human condition have recently come back into favor in some social sciences. Dennett (1987), Hawkins (1997), and Farber (1994) worry that sociobiology is dangerously close to social Darwinism, which they argue has been discredited. Kitcher (1985) deconstructs and severely challenges the methods, theories, and findings of sociobiology.

20. Individualistic lessons supported by Maryanski and Turner (1992). Cooperative lessons supported by Kropotkin (1987).

21. Kellert and Wilson (1993); Kahn (1999).

22. See Shepard (1982).

23. Peterson (2001); Pinker (2002). Also see discussion of humanism earlier in this chapter.

CHAPTER ELEVEN

1. Conditions described by PETA or United Poultry Concerns (http://www.upc-online.org) are not too different than those described in poultry industry publications. Statistics are from the U.S. Department of Agriculture's National Agricultural Statistics Service.

2. Goodwin and Morrison (2002), available with other information at Americans for Medical Progress: http://www.ampef.org.

3. Cosmetics, Toiletry, and Fragrance Association, http://www.ctfa.org.

4. http://www.animalliberationfront.com/ALFront/ALFActs.htm (accessed March 2003). See also a report on ecoterrorism by the Southern Poverty Law Center in their April 2002 Intelligence Report: http://www.splcenter.org/intel/intpro.jsp.

5. Nationwide survey: 47 percent strongly agree and 32 percent moderately agree that animals can be used by humans as long as the animal does not experience undue pain and suffering. In another question, only 11 percent strongly agreed and 10 percent moderately agreed that animals can be utilized regardless of the animal's welfare. Duda, Bissell, and Young (1998, chap. 12).

6. This chapter uses "rights" as a rhetorical tool to capture the broader notion of rights, inherent values, and intrinsic values. Nuanced arguments differentiating among inherent values, intrinsic values, and rights are ignored here in order to focus on the more fundamental questions of how inherent values (or rights) affect understandings of nature and how understandings of nature affect inherent values and rights. Regan (1983) specifically argues for granting rights to all sentient beings. Singer (1975) uses a utilitarian calculus to balance pain and pleasure among all sentient life. Taylor (1986) extends rights and inherent values to parts of the biosphere other than sentient organisms (i.e., species and ecosystems). An accessible and informative debate over these issues can be found in Budiansky (1998) and Scully (2002). Additional perspectives can be found in Callicott (2002), Clark (1997), Midgley (1983), Norton (1987), Preston (2002), and Rolston (1988).

7. The concept of "interests" presents another intellectual thicket that is discussed in more detail later in the chapter. Briefly, interests, as defined here, do not require consciousness. For example, a tree can have an interest in living and not be aware of its interest. I, in contrast, might be interested in the tree staying alive, which requires that I am conscious because it is I who is interested.

8. There exists a long and tortuous history of debate over whether biological entities have inherent rights, and if so, why. Some people believe that certain rights exist as

inherent properties of all biological entities, regardless of human existence, thoughts, or concerns. Secular arguments supporting this position contend that these rights are logically part of any fundamental definition of an entity possessing life; they are *essential* properties of biological life. Rolston (1982, 1988) takes an essentialist position.

Spiritual arguments supporting the existence of inherent rights in nature contend that the rights exist because God deems his creations good or because ancestors and spirits exist within plants and animals. Again there is no way to empirically observe these rights or their assignment. They are assumed to exist, on faith. A third way to define these rights into existence argues that nonhumans depend upon one another for survival. Life is interconnected by the ecological web and all parts of the web are required to sustain life. Thus one part of the web has an interest in the survival of other parts that perform the functions on which all parts depend. Nonhuman organisms, therefore, could be said to provide a source of value for one another, hence providing a nonhuman, nonsupernatural reason for granting rights and value to life on Earth. But, again, we have few means of evaluating or rationally debating the assignment of these rights. Essentialist, spiritual, and codependency positions are hotly debated by religious and ethical philosophers. These debates can seem irrelevant to local people negotiating everyday decisions about their use of land and animals. An alternative, more pragmatic, interests-based approach is presented in this chapter.

Callicott (1989; see esp. chaps. 2, 8, 9) argues that intrinsic values are subjective, requiring a human valuer, and not really open to inspection or negotiation. Norton (1995a, 1995b) argues, pragmatically, that these rights can be negotiated. Preston (1998) examines these three arguments.

9. See Singer (1975). Leopold (1967) argues for extending the moral community to include the "land" in his land ethic essay.

10. Singer (1975) champions equality as an argument for animal rights.

11. VanDeVeer (1979) describes the scheme presented here to evaluate interspecies justice. Taylor (1986) presents a somewhat more elaborate scheme.

12. Bocking (1994), Sagoff (1985), Shrader-Frechette and McCoy (1995). The same argument could be made of some organisms. In the case of certain fungal, coral, and clonal organisms, it is difficult to know where one unit ends and another one begins.

13. Details about the challenges to defining ecological units are provided in Evernden (1992), Fitzsimmons (1999), Sagoff (1985, 1988b), and Shrader-Frechette and McCoy (1995).

14. Stone (1987, 60-62, chaps. 8, 9).

15. Regan (1983) critiques the fascism of ecological ethics such as Leopold (361-63). Callicott (1989) responds to that critique.

16. The system proposed by Norton (1995b) is discussed. For related but alternative systems, see Callicott (1989, chap. 3; 1999, chap. 4).

17. Dizard (1994), Lutts (1990, 198-202), and Nelson (1997) provide worthwhile discussions of deer controversies.

18. Copp (1975, 67).

19. Budiansky (1998, 194).

CHAPTER TWELVE

1. Cordell (1999); Driver, Brown, and Peterson (1991).

2. Mountain biking statistics from Cordell (1999). Stebbins (1982) summarizes the role of "serious leisure" in our lives. Bryan (1977) reviews the commitment, values, and

life satisfaction of specialized recreationists, especially fishers. Details about mountain biking come from interviews I did with members of the Iron Mountain Bike Club in southwest Virginia in 1996.

3. Certainly the idea of leisure predates the Industrial Revolution. The Greeks spoke of leisure as a time of learning and social debate. However, the modern definition of leisure emerged with the Industrial Revolution, which produced a work-leisure dichotomy (Kelly 1992).

4. Huth (1957) and Smith (1950) review early American recreation.

5. Henry Herbert wrote under the pen name Frank Forester in the 1830s. See also William H. H. Murray's *Camplife in the Adirondacks* and Major John M Gould's *Hints for Camping and Walking: How to Camp* (Huth 1957).

6. History taken from Sellars (1997).

7. Wilson (1992, esp. chap. 1, 19-51). See also Hays (1987).

8. Figures taken from Cordell (1999, chap. 5).

9. Originally only the National Park Service was charged with providing recreation on federal lands. The Forest Service, concerned that increasing recreation visits to its lands would be used as an argument by the National Park Service to take responsibility for some Forest Service lands, pressed for the passage of the Multiple-Use Sustained-Yield Act in 1960. This act legitimized recreation along with water, wildlife, timber, and range as multiple uses on forested lands. The Bureau of Land Management followed suit with a similar effort.

10. Cordell (1999).

11. Kaplan and Kaplan (1989) provide numerous examples of research projects that document restorative and related benefits of recreation. See also Driver, Brown, and Peterson (1991) and Driver (1996).

12. Elliot (1997, 81-93) describes the island example as a thought experiment. Kreiger (1973) discusses a similar issue with respect to plastic versus real trees as landscaping. Lack of authenticity may be one reason why exotic species are valued less than native nature.

13. The history of the text that became the Organic Act is reviewed by Sellars (1997, chaps. 1, 2, esp. 38-41).

14. The bear controversy is reviewed by Sellars (1997, 79-80, 160-61, 249-53). See also Chase (1987).

15. Leopold quotes are taken from the "Conservation Ethic" essay in *A Sand County Almanac* (1967, 165-77).

16. Sagoff raises these and other vexing questions about recreation (1974, 205-12). Driver, Brown, and Peterson (1991) describe and defend the benefits of recreation. Abbey (1981) unflinchingly champions authentic, wild, nature-based experiences.

17. Muir (2001, chap. 4).

18. Botkin (1990, 2001) compares classical aesthetics and *The Harvest* to ecological aesthetics. See also Callicott (1992), Nassauer (1992, 1997), and Saito (1998).

19. Parsons and Daniel (2002) defend the biological basis of aesthetic preference.

20. Nassauer (1997).

21. Jenkins (1994, 146).

22. Borman, Balmori, and Geballe (1993); Jenkins (1994).

23. Several landscape philosophers and practitioners advocate the importance of aesthetics and the need to nurture a more ecological or biocultural aesthetic: See Callicott (1992), Gobster (1999), Leopold (1967), Nassauer (1997), and Tuan (1993).

CHAPTER THIRTEEN

1. John Locke (1986, "The Second Treatise of Civil Government": at the beginning of chap. 2, "The State of Nature"). Italics added.

2. Miller (1988); Stoll (1997).

3. Marx (1964, 125), quoting Thomas Jefferson in *Notes on the State of Virginia*; see also Miller (1988).

4. Marx (1964, 124–25), quoting Jefferson's *Notes on the State of Virginia*.

5. Marx (1960, 159), quoting Jefferson, *Notes on the State of Virginia*. Jefferson also revised his critique of mechanization, which when applied in rural settings facilitated working with nature, hence furthering the pastoral ideal. The machine, labor, and nature, then, could combine in America to produce "independent and moral" citizens.

6. Schmitt (1990) calls it Arcadia.

7. Published in *Horticulture* (July 1848); found in Schmitt (1990, 56).

8. Jacob (1997) analyzes these people and their effect on the landscape in a book titled *New Pioneers*. Two examples of the literature and industry supporting rural migration can be found in John Clayton (2001) and Ross and Ross (1997).

9. Macie and Hermansen (2002).

10. Harding and Meyer (1980) provide one of the many biographies of Thoreau.

11. Thoreau (1960, 8–9).

12. Ibid., 71. Interpretation taken from Oelschlaeger (1991, 155) and Turner (1989). Thoreau was embedded in and influenced by a growing critique of industrialization from both sides of the Atlantic (i.e., Karl Marx, Emerson, Carlyle); see Marx (1964, chap. 4).

13. Thoreau (1960, 66).

14. Nash (1982, chap. 5), quoting and interpreting Thoreau.

15. Humans as wild nature grown self-consciousness and the following characterization of Thoreau are offered by Oelschlaeger (1991, 171).

16. The "part and parcel of nature" quote comes from the opening of Thoreau's essay "Walking"; the "Bean Field" is chapter 7 of *Walden*. Thomashow (1995, 32–33) makes a similar argument, only using as an example Thoreau's comparison of his wood furniture to the trees from which they were made.

17. Smith (1950, esp. 53–58).

18. Turner's analysis has been questioned but remains popular. See Nash (1982), Oelschlaeger (1991, 209), and Smith (1950).

19. Schmitt (1990, chap. 7) reviews this history. Quotes are from Mary Dickerson, "Nature Study in City Primary Schools," *Nature Study Review* 2 (March 1906): 100, quoted in Schmitt (1990, 85). Lutts (1990) reviews after-school programs.

20. Quoted in Schmitt (1990, 83–84), who is citing articles published in 1905 and 1907 in the *Nature Study Review*, a scholarly/professional journal that began publishing in 1905 and for twenty years provided a forum for environmental education.

21. Statistics come from Schmitt (1990, chaps. 8, 9).

22. Ibid., chap. 10. The quote is from page 109. Statistics come from Boy Scouts of America fact sheets: http://www.scouting.org/nav/enter.jsp?s=mc&c=fs; and from "About Girl Scouts of the USA": http://www.girlscouts.org/about/#facts (both accessed February 2003).

23. See Orr (1992) for a passionate plea for ecological literacy. The quotes come from pages 87 and 88 of his book. See also the North American Association for Environmental Education: http://naaee.org/aboutnaaee/index.php.

24. More examples and explanations of concerns can be found in Satchell (1996), Sanera and Shaw (1999), and Stuebner (2001).

25. Norton (1987) discusses transformation as a difficult to quantify benefit of nature. See also Leopold, "Conservation Ethic" (1967), and Callicott (1989, chap. 12).

26. Kaplan and Kaplan (1989) review how extended-stay nature camp experiences transform campers' understandings of themselves and of nature.

27. Jackson (1980); Sagoff (1974; 1988); Norton (2003); Tuan (1974).

28. Sagoff (1988a, 63).

29. Evernden (1992, chap. 1) argues that the realization that many of nature's lessons are socially constructed has a paradoxical effect. Many of us seek absolute truths to guide us through difficult decisions in a changing and confusing world. We find it unsettling that all societal norms are human creations and thus subject to endless revision. So we look for moral lessons in the bedrock of nature, in the realm beyond the currents of human culture. We crave certainty in a tide of relativity. The more independent and self-willed we become, then the more we seem tempted to seek an absolute in the contrasting realm of nature.

CONCLUSION

1. Diamond (2005), Posner (2004).

2. Antonousky (1987) provides a compelling example of this basic theory of social psychology using concentration camp survivors and others who have found ways to survive difficult and stressful situations.

3. Curry (2003); Dukes, Piscolish, and Stephens (2000); Dryzek (2000), Lewicki, Gray, and Elliott (2003).

4. Posner (1996) discusses how public participation following the National Environmental Policy Act allows people's concerns to be dismissed as "interests" and favors only "facts" supported by established discourses. See also Curry (2003), Gregory (2002), Satterfield, Slovic, and Gregory (2000), and Satterfield (2001), who talk about the advantages of pluralization and alternative ways to assess and express values for nature.

References

Abbey, Edward. 1981. *Desert Solitaire: A Season in the Wilderness*. Salt Lake City: Peregrine Smith.

Ackerman, Frank, and Lisa Heinzerling. 2004. *Priceless: On Knowing the Price of Everything and the Value of Nothing*. New York: New Press.

Allen, John. 1991. *Biosphere 2: The Human Experiment*. New York: Penguin.

Allen, T. F. H., et al. 2001. "Dragnet Ecology—'Just the Facts, Ma'am': The Privilege of Science in a Postmodern World." *BioScience* 51 (6): 475–85.

Allen, Tim, Joseph Tainter, and Thomas Hoekstra. 2003. *Supply-side Sustainability*. New York: Columbia University Press.

Angermeier, P. L., and J. R. Karr. 1994. "Biological Integrity versus Biological Diversity as Policy Directives." *Bioscience* 44 (10): 690–97.

Antonousky, Aaron. 1987. *Unraveling the Mystery of Health: How People Manage to Stress and Stay Well*. San Francisco: Jossey Bass.

Arrow, K., et al. 1995. "Economic Growth, Carrying Capacity, and the Environment." *Science* 268:520–21.

Aukerman, Dale. 1993. *Reckoning with Apocalypse: Terminal Politics and Christian Hope*. Elgin, IL: Brethren Press.

Bailey, Liberty H. 1917. *The Holy Earth*. New York: Scribner.

Balmford, Andrew, et al. 2002. "Economic Reasons for Conserving Wild Nature." *Science* 297:950–53.

Bannister, Robert C. 1979. *Social Darwinism: Science and Myth in Anglo-American Social Thought*. Philadelphia: Temple University Press.

Barcott, Bruce. 2001. "For God So Loved the World." *Outside* (March): 84–91, 121–24.

Barnett, H., and C. Morse. 1963. *Scarcity and Growth: The Economics of Natural Resource Availability*. Baltimore: Johns Hopkins University Press.

Barrow, Mark. 2002. "Science, Sentiment, and the Specter of Extinction." *Environmental History*, 69–98.

Bast, H. L., P. J. Hill, and R. C. Rue. 1994. *Eco-sanity: A Common-Sense Guide to Environmentalism*. Lanham, MD: Madison Books.

Beatley, Timothy. 1994. *Ethical Land Use: Principles of Policy and Planning*. Baltimore: Johns Hopkins University Press.

Beck, Ulrich. 1992. *Risk Society: Towards a New Modernity*. Translated by M. Ritter. Newbury Park, CA: Sage.

Benyus, Janine. 1998. *Biomimicry: Innovation Inspired by Nature*. New York: William Morrow.

Bernstein, Richard. 1988. "Metaphysics, Critique, and Utopia." *Review of Metaphysics* 42 (2): 255–73.

Berry, Wendell. 1993. "Christianity and the Survival of Creation." In *Sacred Trusts: Essays on Stewardship and Responsibility*, edited by M. Katakis. San Francisco: Mercury House.

Black, John. 1970. *The Dominion of Man: The Search for Ecological Responsibility*. Edinburgh: Edinburgh University Press.

Bocking, Stephen. 1994. "Visions of Nature and Society: A History of the Ecosystem Concept." *Alternatives* 20 (3): 12–18.

———. 1997. *Ecologists and Environmental Policy: A History of Contemporary Ecology*. New Haven, CT: Yale University Press.

Bonnicksen, Thomas M. 2000. *America's Ancient Forests: From the Ice Age to the Age of Discovery*. New York: John Wiley and Sons.

Bookchin, Murray. 1982. *The Ecology of Freedom: The Emergence and Dissolution of Hierarchy*. Palo Alto, CA: Cheshire Books.

———. 1989. *Remaking Society*. Montreal: Black Rose Books.

Bormann, F. H., D. Balmori, and G. Geballe. 1993. *Redesigning the American Lawn: A Search for Environmental Harmony*. New Haven, CT: Yale University Press.

Botkin, Daniel. 1990. *Discordant Harmonies*. London: Oxford University Press.

———. 2001. *No Man's Garden: Thoreau and a New Vision for Civilization and Nature*. Washington, DC: Island Press.

Boulding, Kenneth. 1973. "The Economics of the Coming Spaceship Earth." In *Toward a Steady-State Economy*, edited by H. E. Daly. San Francisco: W. H. Freman.

Boyer, Paul. 1992. *When Time Shall Be No More: Prophecy Belief in Modern American Culture*. Cambridge, MA: Harvard University Press.

Bright, Chris. 1998. *Life Out of Bounds: Bioinvasion in a Borderless World*, Worldwatch Environmental Series Alert. New York: Norton.

Browers, Michaelle L. 1999. "Jefferson's Land Ethic: Environmental Ideas in *Notes on the State of Virginia*." *Environmental Ethics* 21 (Spring): 43–57.

Brown, Lester. 1978. *The Twenty-ninth Day: Accommodating Human Needs and Numbers to the Earth's Resources*. New York: Norton.

Bryan, Hobson. 1977. "Leisure Value Systems and Recreational Specialization: The Case of Trout Fishermen." *Journal of Leisure Research* 9:174–87.

Budiansky, Stephen. 1998. *If a Lion Could Talk: Animal Intelligence and the Evolution of Consciousness*. New York: Free Press.

Bullard, Robert D. 2000. *Dumping in Dixie: Race, Class, and Environmental Quality*. 3rd ed. Boulder, CO: Westview Press.

Callenbach, Ernest. 1975. *Ecotopia: The Novel of Your Future*. New York: Bantam.

Callicott, J. B. 1988. "Animal Liberation and Environmental Ethics: Back Together Again." *Between the Species* 4:163–69.

———. 1989. *In Defense of the Land Ethic: Essays in Environmental Philosophy*. Albany: State University of New York Press.

———. 1992. "The Land Aesthetic." *Renewable Resources Journal* (Winter): 12–17.

————. 1994. *Earth's Insights: A Multicultural Survey of Ecological Ethics from the Mediterranean Basin to the Australian Outback*. Berkeley: University of California Press.

————. 1999. *Beyond the Land Ethic: More Essays in Environmental Philosophy*. Albany: State University of New York Press.

————. 2002. "The Pragmatic Power and Promise of Theoretical Environmental Ethics: Forging a New Discourse." *Environmental Values* 11:3–25.

Callicott, J. B., L. B. Crowder, and K. Mumford. 1999. "Current Normative Concepts in Conservation." *Conservation Biology* 13:22–35.

Calvin, John. 1950. *Institutes of the Christian Religion*. 2 vols. Vol. 1, *The Library of Christian Classics*. Translated by F. Battles. Edited by J. McNeill. Philadelphia: Westminster Press.

Carson, Rachel L. 1962. *Silent Spring*. New York: Houghton Mifflin.

Chase, Alston. 1987. *Playing God in Yellowstone: The Destruction of America's First National Park*. New York: Harcourt Brace.

————. 1995. *In a Dark Wood: The Fight Over Forests and the Rising Tyranny of Ecology*. Boston: Houghton Mifflin.

Cherry, Conrad. 1980. *Nature and Religious Imagination: From Edwards to Bushnell*. Philadelphia: Fortress Press.

Clark, Stephen R. L. 1994. "Global Religion." In *Philosophy and the Natural Environment*, edited by Robin Attfield and Andrew Belsey. Cambridge: Cambridge University Press.

————. 1997. *Animals and Their Moral Standing*. London: Routledge.

Clayton, John. 2001. *Small Town Bound*. Philadelphia: Xlibris.

Cole, Luke W., and Sheila R. Foster. 2001. *From the Ground Up: Environmental Racism and the Rise of the Environmental Justice Movement*. New York: New York University Press.

Collins, J. 1959. "Darwin's Impact on Philosophy." *Thought* 34:185–248.

Copp, John D. 1975. "Why Hunters Like to Hunt." *Psychology Today* (December): 60–62, 67.

Cordell, H. Kenn, ed. 1999. *Outdoor Recreation in American Life: A National Assessment of Demand and Supply Trends*. Champaign, IL: Sagamore.

Costanza, Robert, et al. 1997. "The Value of the World's Ecosystem Services and Natural Capital." *Nature* 387:255–60.

Cronon, William. 1983. *Changes in the Land: Indians, Colonists, and the Ecology of New England*. New York: Hill and Wang.

————, ed. 1995. *Uncommon Ground: Toward Reinventing Nature*. New York: Norton.

Cross, Frank B. 1997. "The Subtle Vices Behind Environmental Values." *Duke Environmental Law and Policy Forum* 8:151–72.

Curry, Patrick. 2003. "Re-thinking Nature: Towards an Eco-pluralism." *Environmental Values* 12:227–360.

Curti, Merle. 1968. *Human Nature in American Historical Thought*. Columbia: University of Missouri Press.

Daily, Gretchen C., ed. 1997. *Nature's Services: Societal Dependence on Natural Ecosystems*. Washington, DC: Island Press.

Daily, Gretchen C., and Katherine Ellison. 2002. *The New Economy of Nature: The Quest to Make Conservation Profitable*. Washington, DC: Island Press.

Dawkins, Marian Stamp. 1998. *Through Our Eyes Only?: The Search for Animal Consciousness*. New York: Oxford University Press.

Dawkins, Richard. 1996. *The Blind Watchmaker: Why Evidence of Evolution Reveals a Universe without Design*. New York: Norton.

Degler, Carl N. 1991. *In Search of Human Nature: The Decline and Revival of Darwinism in American Social Thought.* New York: Oxford University Press.

DeGrazia, David. 1996. *Taking Animals Seriously: Mental Life and Moral Status.* New York: Cambridge University Press.

Denevan, William M. 1992. "The Pristine Myth: The Landscape of the Americas in 1492." *Annals of the Association of American Geographers* 82 (3): 369–85.

Dennett, D. C. 1987. *The Intentional Stance.* Cambridge, MA: MIT University Press.

Diamond, Jared. 1999. *Guns, Germs, and Steel: The Fates of Human Societies.* New York: Norton.

———. 2005. *Collapse: How Societies Choose to Fail or Succeed.* New York: Viking.

Dizard, Jan E. 1994. *Going Wild: Hunting, Animal Rights, and the Contested Meaning of Nature.* Amherst: University of Massachusetts Press.

Donald, Merlin. 1991. *Origins of the Modern Mind: Three States in the Evolution of Culture and Cognition.* Cambridge, MA: Harvard University Press.

Driver, B. L., ed. 1996. *Nature and the Human Spirit: Toward an Expanded Land Management Ethic.* State College, PA: Venture.

Driver, B. L., Perry J. Brown, and George Peterson, eds. 1991. *Benefits of Leisure.* State College, PA: Venture.

Dryzek, John S. 1997. *The Politics of the Earth: Environmental Discourses.* New York: Oxford University Press.

———. 2000. *Deliberative Democracy and Beyond: Liberals, Critics, Contestations.* New York: Oxford University Press.

Duda, Mark D., Steven J Bissell, and Kira C. Young. 1998. *Wildlife and the American Mind.* Harrisonburg, VA: Responsive Management.

Dukes, E. Franklin, Mariana A. Piscolish, and John B. Stephens. 2000. *Reaching for Higher Ground in Conflict Resolution.* San Francisco: Jossey Bass.

Dunlap, Thomas R. 2004. *Faith in Nature: Environmentalism as Religious Quest.* Seattle: University of Washington Press.

Dutney, Andrew. 2001. *Playing God: Ethics and Faith.* Sydney: Harper Collins Religious.

Easterbrook, Greg. 1995. *A Moment on the Earth: The Coming Age of Environmental Optimism.* New York: Viking.

Eden, Sally. 1998. "Environmental Issues: Knowledge, Uncertainty and the Environment." *Progress in Human Geography* 22 (3): 425–32.

Egerton, Frank N. 1973. "Changing Concepts of the Balance of Nature." *Quarterly Review of Biology* 48:322–50.

Ehrenfeld, D. 1981. *The Arrogance of Humanism.* London: Oxford University Press.

Ehrlich, Paul. 1968. *The Population Bomb:* Sierra Club Books.

Eibl-Eibesfeldt, Irenaus. 1989. *Human Ethology.* New York: Aldine de Gruyter.

Elliot, Robert. 1997. *Faking Nature: The Ethics of Environmental Restoration.* New York: Routledge.

Evernden, Neil. 1992. *The Social Creation of Nature.* Baltimore: Johns Hopkins University Press.

Farber, Paul L. 1994. *The Temptations of Evolutionary Ethics.* Berkeley: University of California Press.

———. 2000. *Finding Order in Nature: The Naturalist Tradition form Linnaeus to E. O. Wilson.* Baltimore: Johns Hopkins University Press.

Ferry, Luc. 1995 (1992). *The New Ecological Order.* Translated by C. Volk. Reprint, Chicago: University of Chicago Press.

Fischer, Frank. 2000. *Citizens, Experts, and the Environment*. Durham, NC: Duke University Press.

Fitzsimmons, Allan K. 1999. *Defending Illusions: Federal Protection of Ecosystems*. Lanham, MD: Rowman and Littlefield.

Flader, Susan. 2004. "Building Conservation on the Land: Aldo Leopold and the Tensions of Professionalism and Citizenship." In *Reconstructing Conservation: Finding Common Ground*, edited by B. A. Minteer and R. E. Manning. Washington, DC: Island Press.

Flader, Susan, and J. B. Callicott. 1991. *The River of the Mother of God*. Madison: University of Wisconsin Press.

Flannery, Tim. 2001. *The Eternal Frontier: An Ecological History of North America and Its People*. New York: Atlantic Monthly Press.

Fowler, Robert Booth. 1995. *The Greening of Protestant Thought*. Chapel Hill: University of North Carolina Press.

Frankel, Carl. 1998. *In Earth's Company: Business, Environment, and the Challenge of Sustainability*. Gabriola Island, BC, Canada: New Society.

Freeman, R. Edward, Jessica Pierce, and Richard Dodd. 2000. *Environmentalism and the New Logic of Business*. New York: Oxford University Press.

Freeze, R. Allan. 2000. *The Environmental Pendulum: A Quest for Truth about Toxic Chemicals, Human Health, and Environmental Protection*. Berkeley: University of California Press.

Frome, M. 1966. *Strangers in High Places*. Garden City, NY: Doubleday.

Futuyma, D. J. 1998. *Evolutionary Biology*. 3rd ed. Sunderland, MA: Sinauar Associates.

Gardner, Gary. 2002. *Invoking the Spirit: Religion and Spirituality in the Quest for a Sustainable World* (Worldwatch Paper 164). Washington, DC: Worldwatch Institute.

Gibbs, Lois Marie. 1998. *Love Canal: The Story Continues*. Stony Creek, CT: New Society.

Glacken, C. J. 1967. *Traces on the Rhodian Shore: Nature and Culture in Western Thought from Ancient Times to the End of the Eighteenth Century*. Berkeley: University of California Press.

Gobster, Paul. 1999. "An Ecological Aesthetic for Forest Landscape Management." *Landscape Journal* 18 (1): 54–64.

———. 2005. "Invasive Species as Ecological Threat: Is Restoration an Alternative to Fear-based Resource Management?" *Ecological Restoration* 23 (4).

Gobster, Paul, and R. Bruce Hull, eds. 2000. *Restoring Nature: Perspectives from the Social Sciences and Humanities*. Washington, DC: Island Press.

Gomez-Pompa, Arturo, and Andrea Kaus. 1992. "Taming the Wilderness Myth." *BioScience* 42 (4): 271–79.

Goodenough, Ursual. 1998. *The Sacred Depths of Nature*. New York: Oxford University Press.

Goodin, Robert E., and Simon J. Niemeyer. 2003. "When Does Deliberation Begin? Internal Reflection versus Public Discussion in Deliberative Democracy." *Political Studies* 51:627–49.

Goodwin, Frederick K., and Adrian R. Morrison. 2002. "Why Animal Researchers Must Remember that Human Beings Are Special." *Reason* (March). http://www.amprogress.org/Issues/Issues.cfm?ID=169&c=13.

Gottlieb, Robert. 1993. *Forcing the Spring: The Transformation of the American Environmental Movement*. Washington, DC: Island Press.

Gould, Stephen Jay. 1989. *Wonderful Life: The Burgess Shale and the Nature of History*. New York: Norton.

Graham, John D., and Jonathan B. Wiener, eds. 1995. *Risk versus Risk: Tradeoffs in Protecting Health and the Environment*. Cambridge, MA: Harvard University Press.

Gray, George M., and John D. Graham. 1995. "Regulating Pesticides." In *Risk versus Risk: Tradeoffs in Protecting Health and the Environment*, edited by John D. Graham and Jonathon B. Wiener. Cambridge, MA: Harvard University Press.

Gregory, Robin. 2002. "Incorporating Value Trade-offs into Community-based Environmental Risk Decisions." *Environmental Values* 11 (4): 461–88.

Gregory, Robin, and Paul Slovic. 1997. "A Constructive Approach to Environmental Evaluation." *Ecological Economics* 21:175–81.

Gregory, Robin, and Katharine Wellman. 2001. "Bringing Stakeholder Values into Environmental Policy Choices: A Community-based Estuary Study." *Ecological Economics* 39:37–52.

Grossman, Gene M., and Alan B. Krueger. 1995. "Economic Growth and the Environment." *Quarterly Journal of Economics* 110 (2): 353–77.

Gundersen, Adolf G. 1995. *The Environmental Promise of Democratic Deliberation*. Madison: University of Wisconsin Press.

Gundersen, Lance H., and C. S. Holling, eds. 2002. *Panarchy: Understanding Transformations in Human and Natural Systems*. Washington, DC: Island Press.

Gundersen, Lance H., C. S. Holling, and Stephen Light, eds. 1991. *Barriers and Bridges to the Renewal of Ecosystems and Institutions*. New York: Columbia University Press.

Hajer, Maarten. 1995. *The Politics of Environmental Discourse*. New York: Oxford University Press.

Hardin, G. 1974. "Living on a Lifeboat." *BioScience* 24:561–68.

Hardin, Walter. 1968. "The Tragedy of the Commons." *Science* 162:1243–48.

Harding, Walter, and Michael Meyer. 1980. *The New Thoreau Handbook*. New York: New York University Press.

Harvey, David. 1993. "The Nature of Environment: The Dialectics of Social and Environmental Change." In *Real Problems, False Solutions: Socialist Register 1993*, edited by Ralph Miliband and Leo Panitch. London: Merlin Press.

Hawken, Paul. 1993. *The Ecology of Commerce: A Declaration of Sustainability*. New York: Harper Business.

Hawkins, Mike. 1997. *Social Darwinism in European and American Thought, 1860–1945*. New York: Cambridge University Press.

Hays, S. P. 1959. *Conservation and the Gospel of Efficiency*. Pittsburgh: University of Pittsburgh Press.

———. 1987. *Beauty, Health, and Permanence: Environmental Politics in the United States, 1955–1985*. New York: Cambridge University Press.

Hayward, Tim. 1994. *Ecological Thought: An Introduction*. Cambridge: Polity Press.

Heilbroner, Robert L., and James K. Galbraith. 1990. *The Economic Problem*. 9th ed. Englewood Cliffs, NJ: Prentice-Hall.

Heyes, C. M. 1998. "Theory of Mind in Nonhuman Primates." *Behavioral and Brain Sciences* 21:101–48.

Hoffman, Andrew J. 2000. *Competitive Environmental Strategy: A Guide to the Changing Business Landscape*. Washington, DC: Island Press.

Hornblower, M. 1998. "Next Stop, Home Depot." *Time*, October 19.

Huber, Peter. 1999. *Hard Green: Saving the Environment from the Environmentalists*. New York: Basic Books.

Hull, R. B. et al. 2002. "Assumptions about Ecological Scale and Nature Knowing Best Hiding in Environmental Decisions." *Conservation Ecology* 6 (2): 12. http://www.consecol.org/vol6/iss2/art12.

Hull, R. B., et al. 2003. "Understandings of Environmental Quality: Ambiguities and Values Held by Environmental Professionals." *Environmental Management* 31 (1): 1–13.

Hull, R. B., D. Robertson, and A. Kendra. 2001. "Public Understandings of Nature: A Case Study of Local Knowledge about 'Natural' Forest Conditions." *Society and Natural Resources* 14:325–40.

Huth, Hans. 1957. *Nature and the American*. Lincoln: University of Nebraska Press.

Ingerson, Alice. 1994. "Tracking and Testing the Nature-Culture Dichotomy." In *Historical Ecology: Cultural Knowledge and Changing Landscapes*, edited by C. Crumley. Santa Fe: School of American Research.

Jackson, John B. 1980. *The Necessity for Ruin and Other Topics*. Amherst: University of Massachusetts Press.

Jacob, Jeffrey. 1997. *New Pioneers: The Back-to-the-Land Movement and the Search for a Sustainable Future*. University Park: Pennsylvania State University Press.

Jenkins, Virginia S. 1994. *The Lawn: A History of an American Obsession*. Washington, DC: Smithsonian Institution.

Jordan, William R., III. 2003. *The Sunflower Forest: Ecological Restoration and the New Communion with Nature*. Berkeley: University of California Press.

Kahn, Peter. 1999. *The Human Relationship with Nature: Development and Culture*. Cambridge, MA: MIT University Press.

Kaplan, R., and S. Kaplan. 1989. *The Experience of Nature: A Psychological Perspective*. New York: Cambridge University Press.

Karr, J. R. 2002. "Understanding the Consequences of Human Actions: Indicators from GNP to IBI." In *Just Ecological Integrity: The Ethics of Maintaining Planetary Life*, edited by P. Miller and L. Westra. Lanham, MD: Rowman and Littlefield.

Karr, J. R., and D. R. Dudley. 1981. "Ecological Perspective on Water Quality Goals." *Environmental Management* 5:55–68.

Kaufman, Wallace. 1994. *No Turning Back: Dismantling the Fantasies of Environmental Thinking*. New York: Basic Books.

Kellert, Stephen R. 1996. *The Value of Life: Biological Diversity and Human Society*. Washington, DC: Island Press.

Kellert, Stephen R., and Edward O. Wilson, eds. 1993. *The Biophilia Hypothesis*. Washington, DC: Island Press.

Kelly, John. 1992. *The Sociology of Leisure*. State College, PA: Venture.

Kempton, W., J. S. Boster, and J. A. Hartley. 1995. *Environmental Values in American Culture*. Cambridge, MA: MIT University Press.

Kinzig, A. 2001. "Bridging Disciplinary Divides to Address Environmental and Intellectual Challenges." *Ecosystems* 4:709–15.

Kirk, A. 2001. "Appropriate Technology: The Whole Earth Catalog and Counterculture Environmental Politics." *Environmental History* 6 (3): 374–94.

Kitcher, Philip. 1985. *Vaulting Ambition: Sociobiology and the Quest for Human Nature*. Cambridge, MA: MIT University Press.

Klinkenborg, Verlyn. 2001. "Cow Parts." *Discover* (August): 52–62.

Knudtson, Peter, and David Suzuki. 1992. *Wisdom of the Elders*. New York: Allen and Unwin.

Krech, Shepard. 1999. *The Ecological Indian: Myth and History*. New York: Norton.

Kreiger, Martin. 1973. "What's Wrong with Plastic Trees?" *Science* 179:44–55.

Kropotkin, P. 1987. *Mutual Aid: A Factor of Evolution*. London: Freedom Press.

Kurtz, Paul. 1994. *Toward a New Enlightenment: The Philosophy of Paul Kurtz*. New Brunswick: Transaction.

Lackey, Robert T. 2001. "Values, Policy, and Ecosystem Health." *BioScience* 51:437–43.

Langston, Nancy. 1995. *Forest Dreams, Forest Nightmares: The Paradox of Old Growth in the Inland West*. Seattle: University of Washington Press.

Laroche, Michel, J. Bergeron, and G. Barbaro-Forleo. 2001. "Targeting Consumers Who Are Willing to Pay More for Environmentally Friendly Products." *Journal of Consumer Marketing* 18 (6): 503–18.

LeBlanc, Steven A. 1999. *Prehistoric Warfare in the American Southwest*. Salt Lake City: University of Utah Press.

Lee, K. N. 1993. *Compass and Gyroscope: Integrating Science and Politics for the Environment*. Washington, DC: Island Press.

Lee, Martha F. 1995. *Earth First!: Environmental Apocalypse*. Syracuse, NY: Syracuse University Press.

Lele, Sharachchandra, and Richard B. Norgaard. 1996. "Sustainability and the Scientist's Burden." *Conservation Biology* 10 (2): 354–65.

Lemons, J., ed. 1996. *Scientific Uncertainty and Environmental Problem Solving*. New York: Blackwell Science.

Leopold, Aldo. 1967 (1949). *A Sand County Almanac*. Reprint, New York: Ballantine.

Leopold, Luna B., ed. 1993. *Round River: From the Journals of Aldo Leopold*. Oxford: Oxford University Press.

Levins, Richard, and Richard Lewontin. 1985. *The Dialectical Biologist*. Cambridge, MA: Harvard University Press.

Lewicki, Roy, Barbara Gray, and Michael Elliott. 2003. *Making Sense of Intractable Environmental Conflicts*. Washington, DC: Island Press.

Lewis, M. 1992. *Green Delusions: An Environmentalist Critique of Radical Environmentalism*. Durham, NC: Duke University Press.

Light, Andrew. 2003. "Urban Ecological Citizenship." *Journal of Social Philosophy* 34 (1): 44–63.

Light, Andrew, and Rolston Holmes III. 2003. *Environmental Ethics: An Anthology*. Oxford: Blackwell.

Lilliston, B., and R. Cummins. 1998. "The Organic vs. 'Organic': The Corruption of a Label." *Ecologist* 28(4).

Linner, Bjorn-Ola. 2003. *The Return of Malthus: Environmentalism and Post-war Population-Resource Crises*. Strond, UK: White Horse Press.

Locke, John. 1986 (1690). *The Second Treatise of Civil Government*. Reprint, Buffalo, NY: Prometheus Books.

Lomborg, Bjorn. 2001. *The Skeptical Environmentalist: Measuring the Real State of the World*. New York: Cambridge University Press.

Lovejoy, Arthur O. 1960. *The Great Chain of Being: A Study of the History of an Idea*. New York: Harper & Brothers.

Lovelock, James. 1979. *Gaia: A New Look at Life on Earth*. Oxford: Oxford University Press.

Lowenthal, David. 2000. *George Perkins Marsh: Prophet of Conservation*. Seattle: University of Washington Press.

Luke, Timothy W. 1997. *Ecocritique: Contesting the Politics of Nature, Economy, and Culture*. Minneapolis: University of Minnesota Press.

———. 1999. *Capitalism, Democracy, and Ecology*. Urbana: University of Illinois Press.

Lutts, Ralph H. 1990. *The Nature Fakers: Wildlife, Science, and Sentiment*. Golden, CO: Fulcrum.

Macie, E. A., and L. A. Hermansen. 2002. *Human Influences on Forest Ecosystems: The Southern Wildland-Urban Interface Assessment*. Asheville, NC: USDA Forest Service Southern Research Station, General Technical Report SRS-55.

MacLeish, William. 1994. *The Day Before America: Changing the Nature of a Continent*. Boston: Houghton Mifflin.

MacNaghten, Phil, and John Urry. 1998. *Contested Natures*. London: Sage.

Manes, Christopher. 1990. *Green Rage: Radical Environmentalism and the Unmaking of Civilization*. Boston: Little, Brown.

Mann, Charles C., and Mark L. Plummer. 1995. *Noah's Choices: The Future of Endangered Species*. New York: Knopf.

Marsh, George Perkins. 1885 (1864). *The Earth as Modified by Human Action: A Last Revision of "Man and Nature."* New York: Charles Scribner's Sons.

Martin, Calvin. 1987. *Keepers of the Game: Indian-Animal Relationships and the Fur Trade*. Berkeley: University of California Press.

Marx, Leo. 1964. *The Machine in the Garden: Technology and the Pastoral Ideal in America*. London: Oxford University Press.

Marx, L. 1992. "Environmental Degradation and the Ambiguous Social Role of Science and Technology." *Journal of the History of Biology* 25:449–68.

Maryanski, A., and J. Turner. 1992. *The Social Cage: Human Nature and the Evolution of Society*. Stanford, CA: Stanford University Press.

Mayr, Ernst. 1988. *Towards a New Philosophy of Biology: Observations of an Evolutionist*. Cambridge, MA: Harvard University Press.

McDonough, William, and Michael Braungart. 2002. *Cradle to Cradle: Remaking the Way We Make Things*. New York: North Point Press.

McIntosh, Robert P. 1985. *The Background of Ecology: Concept and Theory*. Cambridge: Cambridge University Press.

McKenzie, Daniel H., D. Eric Hyatt, and V. Janet McDonald, eds. 1992. *Ecological Indicators*. Vol. 1. London: Elsevier Applied Science.

McKibben, Bill. 1989. *The End of Nature*. New York: Anchor Books.

McPherson, E. Gregory. 2003. "Urban Forestry: The Final Frontier?" *Journal of Forestry* 101 (3): 20–25.

McQuillan, Alan. 1993. "Cabbages and Kings: The Ethics and Aesthetics of New Forestry." *Environmental Values* 2:191–222.

Meadows, D. H., et al. 1972. *The Limits to Growth: A Report for the Club of Rome's Project on the Predicament of Mankind*. New York: Universe Books.

Meine, Curt. 1988. *Aldo Leopold: His Life and Works*. Madison: University of Wisconsin Press.

———. 2004. "Conservation and the Progressive Movement: Growing from the Radical Center." In *Reconstructing Conservation: Finding Common Ground*, edited by B. A. Minteer and R. E. Manning. Washington, DC: Island Press.

Menon, A., et al. 1999. "Evolving Paradigm for Environmental Sensitivity in Marketing Programs: A Synthesis of Theory and Practice." *Journal of Marketing Theory and Practice* 7 (2): 1–15.

Merchant, Carolyn. 1989. *Ecological Revolutions: Nature, Gender, and Science in New England*. Chapel Hill: University of North Carolina Press.

Midgley, Mary. 1983. *Animals and Why They Matter*. Athens: University of Georgia Press.

———. 1985. *Evolution as a Religion: Strange Hopes and Stranger Fears*. London: Methuen.

———. 1994. *The Ethical Primate: Humans, Freedom, and Morality*. New York: Routledge.

———. 1995. *Beast and Man: The Roots of Human Nature*. Rev. ed. London: Routledge.

Milanich, J. T., and C. Hudson. 1993. *Hernando de Soto and the Indians of Florida*. Gainesville: University of Florida Press.

Miller, Charles. 1988. *Jefferson and Nature: An Interpretation*. Baltimore: Johns Hopkins University Press.

Miller, Kenneth. 1999. *Finding Darwin's God: A Scientist's Search for Common Ground between God and Evolution*. New York: HarperCollins.

Mink, Claudia G. 1992. *Cahokia City of the Sun: Prehistoric Urban Center in the American Bottom*. Collinsville, IL: Cahokia Mounds Museum Society.

Minteer, Ben A., and Robert E. Manning. 2003. *Reconstructing Conservation: Finding Common Ground*. Washington, DC: Island Press.

Morris, Simon C. 1998. *The Crucible of Creation: The Burgess Shale and the Rise of Animals*. Oxford: Oxford University Press.

Moyers, Bill, and the Center for Investigative Reporting. 1990. *Global Dumping Ground: The International Traffic in Hazardous Waste*. Washington, DC: Seven Locks Press.

Muir, John. 2001. *The Mountains of California*: New York: Modern Library.

Mumford, Lewis. 1961. *The City in History*. New York: Harcourt, Brace & World.

Murdoch, W., and J. Connell. 1973. "All about Ecology." In *Western Man and His Environment*, edited by I. Barbour. Reading, MA: Addison-Wesley.

Myers, N. 1982. "Room in the Ark?" *Bulletin of the Atomic Scientists* (November): 44–48.

Myers, Nancy. 2002. "The Precautionary Principle Put Values First." *Bulletin of Science, Technology and Society* 22 (3): 210–19.

Nash, Roderick Frazier. 1982. *Wilderness and the American Mind*. New Haven, CT: Yale University Press.

Nassauer, Joan Iverson. 1992. "The Appearance of Ecological Systems as a Matter of Policy." *Landscape Ecology* 6 (4): 239–50.

———. 1995. "Messy Ecosystems, Orderly Frames." *Landscape Journal* 14 (2): 161–70.

———, ed. 1997. *Placing Nature: Culture and Landscape Ecology*. Washington, DC: Island Press.

Nelson, Richard. 1997. *Heart and Blood: Living with Deer in America*. New York: Vintage.

Nelson, Robert. 2000. *A Burning Issue: A Case for Abolishing the US Forest Service*. New York: Rowan and Littlefield.

Ness, Arne. 1973. "The Shallow and the Deep, Long-Range Ecology Movements: A Summary." *Inquiry* 16 (spring): 95–100.

Newton, J. L., and E. T. Freyfogle. 2005. "Sustainability: A Dissent." *Conservation Biology* 19 (1): 23–32.

Norton, B. G. 1987. *Why Preserve Natural Variability*. Princeton, NJ: Princeton University Press.

———. 1988. "The Constancy of Leopold's Land Ethic." *Conservation Biology* 2:93–102.

———. 1989. "Intergenerational Equity and Environmental Decisions: A Model Using Rawls' Veil of Ignorance." *Ecological Economics* 1 (2): 137–59.

———. 1990. "Context and Hierarchy in Aldo Leopold's Theory of Environmental Management." *Ecological Economics* 2:119–27.

———. 1991. *Toward Unity Among Environmentalists*. New York: Oxford University Press.

———. 1993. "Should Environmentalists Be Organicists?" *Topoi* 12:21–30.

———. 1995a. "Caring for Nature: A Broader Look at Animal Stewardship." In *Ethics on the Ark: Zoos, Animal Welfare, and Wildlife Conservation*, edited by B. Norton et al. Washington, DC: Smithsonian Press.

————. 1995b. "Ecological Integrity and Social Values: At What Scale?" *Ecosystem Health* 1 (4): 228–41.

————. 1995c. "Why I Am Not a Nonanthropocentrist: Callicott and the Failure of Monistic Inherentism." *Environmental Ethics* 17:341–58.

————. 1998a. "Evaluation and Ecosystem Management: New Directions Needed?" *Landscape and Urban Planning* 40:185–94.

————. 1998b. "Improving Ecological Communication: The Role of Ecologists in Environmental Policy Formation." *Ecological Applications* 8 (2): 350–64.

————. 2003. *Searching for Sustainability: Interdisciplinary Essays in the Philosophy of Conservation Biology.* New York: Cambridge University Press.

————. 2005. *Sustainability.* Chicago: University of Chicago Press.

O'Brien, Mary H. 1993. "Being a Scientist Means Taking Sides." *BioScience* 43 (10): 706–8.

Oelschlaeger, Max. 1991. *The Idea of Wilderness.* New Haven, CT: Yale University Press.

————. 1994. *Caring for Creation: An Ecumenical Approach to the Environmental Crisis.* New Haven, CT: Yale University Press.

Opie, John. 1998. *Nature's Nation: An Environmental History of the United States.* Fort Worth, TX: Harcourt Brace College Publishers.

Orr, David. 1992. *Ecological Literacy: Education and the Transition to a Postmodern World.* Albany: State University of New York.

————. 2004. *The Last Refuge: Patriotism, Politics, and the Environment in an Age of Terror.* Washington, DC: Island Press.

Ottman, Jacquelyn A. 1998. *Green Marketing: Opportunity for Innovation.* Chicago: NTC Business Books.

Parsons, R., et al. 1998. "The View from the Road: Implications for Stress Recovery and Immunization." *Journal of Environmental Psychology* 18:113–40.

Parsons, Russ, and T. C. Daniel. 2002. "Good Looking: In Defense of Scenic Landscape Aesthetics." *Landscape and Urban Planning* 60:43–56.

Passmore, John. 1980. *Man's Responsibility for Nature.* 2nd ed. London: Duckworth.

Peterson, Anna L. 2001. *Being Human: Ethics, Environment, and Our Place in the World.* Berkeley: University of California Press.

Peterson, Tarla R. 1995. "Rooted in the Soil: How Understanding the Perspectives of Landowners Can Enhance the Management of Environmental Disputes." *Quarterly Journal of Speech* 81 (2): 139–66.

————. 1997. *Sharing the Earth: The Rhetoric of Sustainable Development.* Columbia: University of South Carolina Press.

Philp, Richard. 1995. *Environmental Hazards and Human Health.* Boca Raton, FL: Lewis.

Pickett, Steward T. A., Jurek Kolasa, and Clive G. Jones. 1996. *Ecological Understanding: The Nature of Theory and the Theory of Nature.* San Diego: Academic Press.

Pinchot, Gifford. 1910. *The Fight for Conservation.* New York: Doubleday.

Pinker, Steven. 2002. *The Blank Slate: The Modern Denial of Human Nature.* New York: Viking.

Pollan, Michael. 2001. *The Botany of Desire: A Plant's-eye View of the World.* New York: Random House.

————. 2002a. "An Animal's Place." *New York Times Magazine* (November 10): 59–64, 100–101.

————. 2002b. "Power Steer." *New York Times Magazine* (March 31).

Porritt, Jonathon. 1985. *Seeing Green: The Politics of Ecology Explained.* New York: Basil Blackwell.

Posner, J. 1996. "A Civic Republican Perspective on the National Policy Act's Process for Citizen Participation." *Environmental Law* 26:53–94.

Posner, Richard A. 2004. *Catastrophe: Risk and Response*. New York: Oxford University Press.

Postman, Neil. 1985. *Amusing Ourselves to Death: Public Discourse in the Age of Show Business*. New York: Viking.

Powers, C. W., and M. R. Chertow. 1997. "Industrial Ecology." In *Thinking Ecologically: The Next Generation of Environmental Policy*, edited by M. Chertow and D. Esty. New Haven, CT: Yale University Press.

Premack, D., and A. Premack. 2003. *Original Intelligence*. New York: McGraw-Hill.

Presbyterian Church. 1990. *Restoring Creation for Ecology and Justice*. Adopted by the 202nd General Assembly. Louisville, KY: Presbyterian Church (USA).

———. 1996. *Hope for a Global Future: Towards Just and Sustainable Human Development*: Approved by the 208th General Assembly. Louisville, KY: Presbyterian Church (USA).

Preston, Christopher J. 1998. "Epistemology and Intrinsic Values: Norton and Callicot's Critiques of Rolston." *Environmental Ethics* 20:409–28.

———. 2002. "Animality and Morality: Human Reason as an Animal Activity." *Environmental Values* 11:427–42.

Price, Jennifer. 1999. *Flight Maps: Adventures with Nature in Modern America*. New York: Basic Books.

Prothero, Andrea, and J. A. Fitchett. 2000. "Greening Capitalism: Opportunities for a Green Community." *Journal of Macromarketing* 20 (1): 46–55.

Pyne, Stephen J. 2004. *Tending Fire: Coping with America's Wildland Fires*. Washington, DC: Island Press.

Redman, Charles L. 1999. *Human Impact on Ancient Environments*. Tucson: University of Arizona Press.

Regan, Thomas. 1983. *The Case for Animal Rights*. Berkeley: University of California Press.

Ridley, Matt. 2003. *Nature via Nurture: Genes, Experiences, and What Makes Us Human*. New York: HarperCollins.

Robertson, D., and R. Bruce Hull. 2003a. "Biocultural Ecology: Exploring the Social Construction of the Southern Appalachian Ecosystem." *Natural Areas Journal* 23 (2): 180–89.

———. 2003b. "Public Ecology: An Environmental Science and Policy for a Global Society." *Environmental Science and Policy* 6:399–410.

Rollin, Bernard E. 1989. *The Unheeded Cry: Animal Consciousness, Animal Pain, and Science*. New York: Oxford University Press.

Rolston, Holmes, III. 1981. "Values in Nature." *Environmental Ethics* 3:113–28.

———. 1982. "Are Values in Nature Subjective or Objective?" *Environmental Ethics* 4:125–51.

———. 1988. *Environmental Ethics*. Philadelphia: Temple University Press.

Romm, Joseph. 1994. *Lean and Clean Management: How to Boost Profits and Productivity by Reducing Pollution*. New York: Kodansha International.

Roosevelt, Teddy. 1910. *Hunting Trips on the Prairie and in the Mountains*. New York: Review of Reviews.

Rosenzweig, Michael. 2003. *Win-Win Ecology: How the Earth's Species Can Survive in the Midst of Human Enterprise*. New York: Oxford University Press.

Ross, J. P. 2002. "Sempra: Exporting Pollution: U.S.-Mexico Border Region to Pay the Price for California's Power." *CorpWatch* (May 27).

Ross, Marilyn, and Tom Ross. 1997. *Country Bound!: Trading Your Business Suit Blues for Blue Jean Dreams*. Denver: UpStart.

Rossing, Barbara R. 2004. *The Rapture Exposed: The Message of Hope in the Book of Revelation*. Boulder, CO: Westview.

Rothman, Dale S. 1998. "Environmental Kuznets Curves—Real Progress or Passing the Buck? A Case for Consumption-based Approaches." *Ecological Economics* 25: 177–94.

Sagoff, Mark. 1974. "On Preserving the Natural Environment." *Yale Law Review* 81:205–67.

———. 1981. "At the Shrine of Our Lady of Fatima or Why Political Questions Are Not All Economic." *Arizona Law Review* 23:1283–98.

———. 1985. "Fact and Value in Ecological Science." *Environmental Ethics* 7 (2): 99–116.

———. 1988a. *The Economy of the Earth*. Cambridge: Cambridge University Press.

———. 1988b. "Ethics, Ecology, and the Environment: Integrating Science and Law." *Tennessee Law Review* 56:77–229.

———. 1997. "Can We Put a Price on Nature's Services?" *Report from the Institute for Philosophy and Public Policy, School of Public Affairs, University of Maryland* 17 (3): 7–12.

———. 1999. "What's Wrong with Exotic Species?" *Report from the Institute for Philosophy and Public Policy, School of Public Affairs, University of Maryland* 19 (4): 16–23.

———. 2004. *Price, Principle, and the Environment*. Cambridge: Cambridge University Press.

Saito, Y. 1998. "The Aesthetics of Unscenic Nature." *Journal of Aesthetic and Art Criticism* 56:101–11.

Sale, Kirkpatrick. 1985. *Dwellers in the Land: The Bioregional Vision*. San Francisco: Sierra Club Books.

Sanera, Michael, and Jane S. Shaw. 1999. *Facts Not Fear: Teaching Children about the Environment*. Washington, DC: Regnery.

Satchell, Michael. 1996. "Dangerous Waters? Why Environmental Education Is Under Attack in the Nation's Schools." *U.S. News and World Report*, June 10, 63–64.

Satterfield, Terre, Paul Slovic, and Robin Gregory. 2000. "Narrative Valuation in a Policy Judgment Context." *Ecological Economics* 34:315–31.

Satterfield, Theresa. 2001. "In Search of Value Literacy: Suggestions for the Elicitation of Environmental Values." *Environmental Values* 10:331–59.

Scarce, R. 2000. *Fishy Business: Salmon, Biology, and the Social Constructions of Nature*. Philadelphia: Temple University Press.

Scherer, Glenn. 2004. "The Godly Must Be Crazy: Christian-Right Views Are Swaying Politicians and Threatening the Environment." *Grist Magazine: A Beacon in the Smog*, October 27. http://www.grist.org.

Schmidheiny, Stephan, with the Business Council for Sustainable Development. 1992. *Changing Course: A Global Business Perspective on Development and the Environment*. Cambridge, MA: MIT Press.

Schmidheiny, Stephan, Federico Zorraquin, and World Business Council for Sustainable Development. 1996. *Financing Change: The Financial Community, Eco-Efficiency, and Sustainable Development*. Cambridge, MA: MIT University Press.

Schmitt, Peter J. 1990 (1969). *Back to Nature: The Arcadian Myth in Urban America*. Reprint, Baltimore: Johns Hopkins University Press.

Schon, Donald A., and Martin Rein. 1994. *Frame Reflection: Toward the Resolution of Intractable Policy Controversies*. New York: Basic Books.

Scully, Matthew. 2002. *Dominion: The Power of Man, the Suffering of Animals, and Call to Mercy*. New York: St. Martin's Press.

Sears, P. B. 1964. "Ecology—a Subversive Subject." *BioScience* 14:11-13.

Sedjo, Roger. 1995. "Forests—Conflicting Signals." In *The True State of the Planet*, ed by R. Bailey. New York: Free Press.

Sellars, Richard. 1997. *Preserving Nature in National Parks: A History*. New Haven, Yale University Press.

Senecah, S. 1996. "Forever Wild or Forever in Battle: Metaphors of Empowermen the Continuing Controversy Over the Adirondacks." In *Earthtalk: Communica Empowerment for Environmental Action*, edited by Star A. Muir and Thomas Veenendall. Westport, CT: Praeger.

Seton, Ernest Thompson. 1907. "The Natural History of the Ten Commandmen *Century Illustrated Monthly Magazine* 75 (November): 24-33.

Shellenberger, Michael, and Ted Nordhaus. 2004. "The Death of Environmentali Global Warming Politics in a Post-Environmental World." http.www.grist.org/ne maindish/2005/01/13/doe-reprint.

Shepard, Paul. 1982. *Nature and Madness*. San Francisco: Sierra Club Books.

Shepard, Paul, and D. McKinley, eds. 1969. *The Subversive Science: Essays toward Ecology of Man*. New York: Houghton Mifflin.

Shettleworth, Sara J. 1998. *Cognition, Evolution, and Behavior*. Oxford: Oxford Univers Press.

Short, John Renne. 1991. *Imagined Country: Environment, Culture, and Society*. Lond Routledge.

Shrader-Frechette, K. 1994. "An Apologia for Activism: Global Responsibility, Ethi Advocacy, and Environmental Problems." In *Ethics and Environmental Policy*, edi by Frederick Ferre and Peter Hartel. Athens: University of Georgia Press.

———. 1995. "Hard Ecology, Soft Ecology, and Ecosystem Integrity." In *Perspectives Ecological Integrity*, edited by L. Westra and J. Lemons. Dordrecht: Kluwer Acader Publishers.

Shrader-Frechette, Kristin S., and Earl D. McCoy. 1995. "Natural Landscapes, Natu Communities, and Natural Ecosystems." *Forest and Conservation History* 39 (138-42.

Shutkin, William. 2000. *The Land That Could Be: Environmentalism and Democracy in t Twenty-first Century*. Cambridge, MA: MIT University Press.

Simmons, I. G. 1989. *Changing the Face of the Earth: Culture, Environment, History*. N York: Blackwell.

Simon, Julian. 1981. *The Ultimate Resource*. Princeton, NJ: Princeton University Press.

Singer, Peter. 1975. *Animal Liberation: A New Ethic for Our Treatment of Animals*. Ne York: New York Review (Random House).

Skinner, B. F. 1948. *Walden Two*. New York: Macmillan.

Slobodchikoff, C. N. 2002. "Cognition and Communication in Prairie Dogs." In *T Cognitive Animal: Empirical and Theoretical Perspectives on Animal Cognition*, edite by M. Bekoff, C. Allen, and G. Burghardt. Cambridge, MA: MIT University Press.

Smith, Bruce. 1998. *The Emergence of Agriculture*. New York: Scientific American Librar

Smith, Henry Nash. 1950. *Virgin Land: The American West as Symbol and Myth*. Cambridg MA: Harvard University Press.

Smith, Michael E. 1996. *The Aztecs*. Oxford: Blackwell.

Smith, Mick. 1999. "To Speak of Trees: Social Constructivism, Environmental Value and the Future of Deep Ecology." *Environmental Ethics* 21 (Winter): 259-376.

Smith, Sherry. 2000. *Reimagining Indians: Native Americans through Anglo Eyes, 1880-1940*. New York: Oxford University Press.

Snyder, G. 1998. "Is Nature Real?" *Resurgence* 190:32.

Soper, Kate. 1995. *What Is Nature?* Cambridge: Blackwell.

Soule, Michael E., and Gary Lease, eds. 1995. *Reinventing Nature? Responses to Postmodern Deconstruction*. Washington, DC: Island Press.

Stebbins, Robert A. 1982. "Serious Leisure: A Conceptual Statement." *Pacific Sociological Review* 25 (2): 251-72.

Steinberg, Ted. 2002. *Down to Earth: Nature's Role in American History*. New York: Oxford University Press.

Stephenson, Kurt. 2003. "An Institutionalist Approach to Environmental Goal Setting." In *Institutional Analysis and Economic Policy*, edited by M. Tool and P. D. Bush. Boston: Kluwer Academic Publishers.

Stoll, M. 1997. *Protestantism, Capitalism, and Nature in America*. Albuquerque: University of New Mexico Press.

———. 2001. "Green versus Green: Religions, Ethics and the Bookchin-Foreman Dispute." *Environmental History Review* 6 (3): 412-27.

Stone, Christopher D. 1987. *Earth and Other Ethics: A Case for Moral Pluralism*. New York: Harper & Row.

Stuebner, Stephen. 2001. "Teach the Children Well: Corporations, Conservationists Vie for Students' Minds in the Unregulated World of Environmental Education." *High Country News*, March 26. http://www.hcn.org.

Sunstein, Cass R. 2002. "The Law of Group Polarization." *Journal of Political Philosophy* 10 (2): 175-95.

———. 2003. "The Rights of Animals." *University of Chicago Law Review* 70 (1): 387-402.

Sutter, Paul S. 2002. *Driven Wild: How the Fight against Automobiles Launched the Modern Wilderness Movement*. Seattle: University of Washington Press.

Swartz, Daniel. 2002. *Jews, Jewish Texts, and Nature: A Brief History*. Coalition on the Environment and Jewish Life. http://www.coejl.org/learn/je_swartz.php.

Takacs, D. 1996. *The Idea of Biodiversity: Philosophies of Paradise*. Baltimore: Johns Hopkins University Press.

Taylor, Bron. 1999. "Green Apocalypticism: Understanding Disaster in the Radical Environmental World." *Society & Natural Resources* 12:377-86.

———. 2004. "A Green Future for Religion?" *Futures* 36:991-1008.

Taylor, Paul W. 1981. "The Ethics of Respect for Nature." *Environmental Ethics* 3:197-218.

———. 1986. *Respect for Nature*: Princeton, NJ: Princeton University Press.

Terrace, Herbert S. 1979. *Nim*. New York: Knopf.

Terrace, Herbert S., et al. 1979. "Can Apes Create a Sentence?" *Science* 206:891-902.

Terrie, Philip G. 1997. *Contested Terrain: A New History of Nature and People in the Adirondacks*. Syracuse, NY: Adirondack Museum/Syracuse University Press.

Thayer, Robert L. 1994. *Gray World, Green Heart: Technology, Nature, and the Sustainable Landscape*. New York: John Wiley and Sons.

Thomas, Keith. 1983. *Man and the Natural World: A History of the Modern Sensibility*. New York: Pantheon.

Thomashow, Mitchell. 1995. *Ecological Identity: Becoming a Reflective Environmentalist*. Cambridge, MA: MIT University Press.

Thoreau, Henry David. 1960. *Walden; or, Life in the Woods*. New York: Penguin.

Throgmorton, J. A. 1991. "The Rhetorics of Policy Analysis." *Policy Sciences* 24:153-79.

Tierney, John. 1996. "Recycling Is Garbage." *New York Times*, June 30, 24-29, 44, 48, 51, 53.

Tomasello, Michael. 1999. *The Cultural Origins of Human Cognition*. Cambridge, MA: Harvard University Press.

Tuan, Yi-Fu. 1974. *Topophilia: A Study of Environmental Perception, Attitudes, and Values*. New York: Columbia University Press.

———. 1993. *Passing, Strange, and Wonderful: Aesthetics, Nature and Culture*. Washington, DC: Island Press.

Turner, Frederick. 1985a. "Cultivating the American Garden: Toward a Secular View of Nature." *Harper's Magazine* (August): 45-52.

———. 1985b. *John Muir: Rediscovering America*. Cambridge, MA: Perseus.

———. 1989. *Spirit of Place: The Making of an American Literary Landscape*. Washington, DC: Island Press.

———. 1994. "The Invented Landscape." In *Beyond Preservation: Restoring and Inventing Landscapes*, edited by Judith De Luce, A. Dwight Baldwin Jr., and Carl Pletsch. Minneapolis: University of Minnesota Press.

Ulrich, R. S. 1984. "View through a Window May Influence Recovery from Surgery." *Science* 224:420-21.

———. 1993. "Biophilia, Biophobia, and Natural Landscapes." In *The Biophilia Hypothesis*, edited by S. R. Kellert and E. O. Wilson. Washington, DC: Island Press.

VanDeVeer, Donald. 1979. "Interspecific Justice." *Inquiry* 22:55-70.

Van Tilburg, Jo Anne. 1994. *Easter Island: Archaeology, Ecology, and Culture*. Washington, DC: Smithsonian Institution Press.

Vaux, Henry J. 2000. *Watershed Management for Potable Water Supply: Assessing the New York City Strategy*. Washington, DC: National Research Council.

Vitousek, Peter M., et al. 1997. "Human Domination of Earth's Ecosystems." *Science* 277 (July 25): 494-99.

Waal, Frans de. 1986. "Deception in the Natural Communication of Chimpanzees." In *Deception: Perspectives on Human and Nonhuman Deceit*, edited by R. W. Mitchell and N. S. Thompson. Albany: SUNY Press.

Wallman, Joel. 1992. *Aping Language*. New York: Cambridge University Press.

Weaver, Bruce. 1996. "What to Do with Mountain People? The Darker Side of the Successful Campaign to Establish the Great Smoky Mountains National Park." In *The Symbolic Earth*, edited by J. Cantrill and C. L. Oravec. Lexington: University Press of Kentucky.

Weber, Edward P. 2000. "A New Vanguard for the Environment: Grass-roots Ecosystem Management as a New Environmental Movement." *Society & Natural Resources* 13: 237-59.

Weber, Max. 1930. *The Protestant Ethic and the Spirit of Capitalism*." Translated by T. Parsons. London: Allen & Unwin.

Westra, L., and J. Lemons, eds. 1995. *Perspectives on Ecological Integrity*. Dordrecht: Kluwer Academic Publishers.

White, Lynn. 1967. "The Historical Roots of Our Ecological Crisis." *Science* 155:1203-7.

White, Richard. 1984. "Native Americans and the Environment." In *Scholars and the Indian Experience*, edited by W. R. Swagerty. Bloomington: Indiana University Press.

———. 1995. *The Organic Machine: The Remaking of the Columbia River*. New York: Hill and Wang.

Whiten, A., and R. W. Byrne. 1988. "Tactical Deception in Primates." *Behavioral and Brain Sciences* 11:223-73.

Wicklum, D., and Ronald W. Davies. 1995. "Ecosystem Health and Integrity?" *Canadian Journal of Botany* 73:997-1000.

Wiener, Jonathan Baert. 1996. "Beyond the Balance of Nature." *Duke Environmental Law and Policy Forum* 7 (1): 1-24.

Wilson, Alexander. 1991. *The Culture of Nature: North American Landscape from Disney to the Exxon Valdez.* Toronto: Between the Lines.

Wilson, Edward O. 1978. *On Human Nature.* Cambridge, MA: Harvard University Press.

———. 1992. *The Diversity of Life.* Cambridge, MA: Belknap Press of Harvard University Press.

Woods, Mark, and Paul V. Moriarty. 2001. "Strangers in a Strange Land: The Problem of Exotics." *Environmental Values* 10:163-91.

World Commission for Environment and Development. 1987. *Our Common Future.* Oxford: Oxford University Press.

Worster, Donald. 1979. *Dust Bowl: The Southern Plains in the 1930s.* New York: Oxford University Press.

———. 1994. *Nature's Economy: A History of Ecological Ideas.* Cambridge: Cambridge University Press.

———. 1996. "The Two Cultures Revisited: Environmental History and the Environmental Sciences." *Environment and History* 2:3-14.

Wright, P. A., et al. 2002. "Monitoring for Forest Management Unit Scale Sustainability: The Local Unit Criteria and Indicators Development (LUCID)." Fort Collins, CO: USDA Forest Service.

Wright, Robert. 2000. *Nonzero: The Logic of Human Destiny.* New York: Pantheon.

Wu, J., and O. L. Loucks. 1995. "From Balance of Nature to Hierarchical Patch Dynamics: A Paradigm Shift in Ecology." *Quarterly Review of Biology* 70:439-66.

Wynne, B. 1992. "Misunderstood Misunderstanding: Social Identities in Public Uptake of Science." *Public Understanding of Science* 1:281-304.

Yearley, Steven. 1996. *Sociology, Environmentalism, Globalization.* London: Sage.

———. 2000. "Making Systematic Sense of Public Discontents with Expert Knowledge: Two Analytical Approaches and a Case Study." *Public Understanding of Science* 9:105-22.

Zimmerer, Karl S. 1994. "Human Geography and the 'New Ecology': The Prospect and Promise of Integration." *Annals of the Association of American Geographers* 84 (1): 108-25.

Index

ecosystem, 41, 52, 172; rights (*see* ecocentric)
ecosystem services. *See* nature's services
ecoterrorism, 55, 230
education, environmental, 64, 175, 200–204
Edwards, Jonathan, 131
Emerson, Ralph Waldo, 131–32, 197
emotional control, 205
endangered species, 32, 170, 205
Enlightenment, 73, 76, 134, 150–51, 192
entertainment, 182–87
environmentalism: death of, 1, 7, 208; join with religion, 138–40; radical, 55–56, 138, 160; shift to equity, 128–29; shift to health, 97–99
environmental justice, 70, 115–16, 129
Environmental Kuznets Curve, 82–83
Environmental Protection Agency (EPA), 88, 92, 107, 110, 112–14
equality and animal rights, 167, 231
equity: displacement through preservation, 118–19; intergenerational, 229–33; racism, 116–17; toxins, 114–16
Erie, Lake, 100
eugenics, 35, 72
eukaryotic cells, 27–28
Evangelical Environmental Network, 139
evolution: cooperative and competitive, defined, 24–25; random, 27–28
evolutionary psychology, 4, 154
exotic species, 17, 21, 46–48, 171–72
extend rights beyond humans, 163–64, 177; Aldo Leopold, 50–53
extinction, 60, 103, 130, 209
exurbia, 196

fairness. *See* equity
farm, 142–43, 173, 180, 193
farm aid, 195
fascism, fascist, 52, 55, 171–72
forestry, 17, 43–44, 49, 69, 91, 139
Franklin, Benjamin, 180
Fresh Air charities, 106, 202
frontier, 58, 75, 122, 186, 195, 200–201
functionalists, 46–48
fundamentalism, 1, 4–5, 48, 171, 208
future generations, 65, 83, 95, 118–22, 172, 181, 206

Gandhi, Mahatma, 162, 200
gardening, 178–79
Garden of Eden, 126, 134, 139
General Grant (tree), 5–6, 162
Genesis, 20, 124–28, 142
genetic engineering, 65, 72, 81, 174
Germany, animal rights, 161
Girl Scouts, 202–3
glacier, 22–24
global economy, 82–83, 94, 114, 208
God's Greens, 129, 139
golden rule, 129
golf, 180, 185, 190
Gondwanaland, 21
great chain of being, 150
Great Smoky Mountains National Park, 117–19, 179
green business, 91–95
green consumerism, 87–91
green technology, 67–68
Grinnell, George Bird, 14

Hamilton, Alexander, 194
hard green, 67
Hetch Hetchy, Yosemite, 132
hierarchy. *See* dynamic equilibrium ecology
Highway Beautification Act, 132
hillbilly, 118
holism model of ecology, 40–41, 44, 45, 50, 55
Hollywood, 14, 138
Home Depot, 93
Homestead Act, 77
Homo sapiens, 22, 30, 32
hubris, 61, 72–73, 128, 215
Hudson Bay, 22
humane animal treatment, 92, 161, 172, 176. *See also* rights
humanism, 151
humanity's fall from glory, 142
human-nature dichotomy, 51, 122, 176–77, 199
hunting: aboriginal, 12, 15, 24; animal rights, 169–70; fascism, 52, 171; sport, 135, 174–75, 180, 182, 204

ice age, 21–24
identity: cultural, 206; recreation, 178–81; self, 192, 204–5

industrialization, Industrial Revolution, 14, 58–60, 75–76, 98, 104–5, 135, 179, 194
inherent value, 161. *See also* rights
injustice. *See* equity
insurance industry, 92
intelligent design, 124, 130, 134
interests that matter, 167–68, 176
intrinsic value, 161. *See also* rights
instrumental value, 161–62, 170, 176
invasive species, 17, 42, 49, 171
is-ought naturalistic fallacy, 4, 34, 38

Jefferson, Thomas, 132, 193–95
John Paul II (pope), 129
Judeo-Christian environmental ethics, 124–40

Kant, Immanuel, 229
Kassa Island, 115
Kettleman City, CA, 116

labeling environmental products, 89
land conversion, 80–81, 96
language abilities of animals, 144–47, 152–53
lawns, 190–91
Leave No Trace organization, 204
Left Behind series (LaHaye and Jenkins), 136
leisure, 179–80
Leopold, Aldo, 3, 49–53, 186
Lewis and Clark, 60, 76
life-cycle product analysis, 89–91
lily pond, 57
limits to growth, 61
Lincoln, Abraham, 77
Lindsey, Hal, 136
Linnaean species classification, 31
livestock, 79–80, 161, 173–74
Locke, John, 192
Love Canal, NY, 100–101
Luther, Martin, 138

Malthus, Thomas, 59, 61, 63
Manifest Destiny, 74–77
Marsh, George Perkins, 59
Massachusetts Missionary magazine, 180
match-to-sample test, 147
Mather, Cotton, 135
McDonald's, 92

mechanistic (model of ecology), 39–40, 42–44, 46, 50
medical research, 71, 159, 161, 169
mirror-recognition test, 47
Mississippi River, 10–11, 22, 49, 76, 100
modernity, modernism, 19, 130, 150, 156, 200
moral decay, 200
moral mirror, 205–6
moral restraint, 204–5
mountain biking, 178
Muir, John: aesthetic, 188; spiritual, 130–32, 200
Multiple-Use Sustained-Yield Act, 181

National Association of Conservation Districts, 140
National Council of Churches, 140
National Environmental Policy Act, 182
National Recreation and Park Association, 106–7
National Trails System Act, 181
National Wild and Scenic Rivers Act, 181
Native Americans: Henry David Thoreau and, 199; Hollywood version, 14; land ethic of, 13–16, 127; land management of, 10–13; tests of character, 200
native species, 44, 47–48
natural capital, 120
naturalistic fallacy, 4, 34, 38
natural regulation, 185
natural selection, 24–25. *See also* evolution
natural theology, 132–33
Nature Conservancy, 54, 139
nature's services: defined and priced, 84–87; substitutes, 66, 211; sustainability, 120, threat of exotics, 48
nature vs. nurture, 154–58. *See also* biophilia
Nazism, 34
neatness, 190
Newton, Isaac, 130, 204
New York City water, 85
NIMBY (not in my backyard), 116
Noah's Ark, 126
noble savage, 13–16
North American Coalition for Christianity and Ecology, 139

Occupational Safety and Health Act, 110
old-growth forest, 44
Olmstead, Frederick Law, 105
organic food, 90, 102, 173-74
organismicism. *See* holism
original nature, 24, 39, 55
oxygen, 21

Pacific Northwest, 60, 120
pain: animal rights, 161-62, 165-69; species comparisons, 144-45
panarchy. *See* dynamic equilibrium ecology
Pandora's box, 71, 100
passenger pigeon, 60
pastoralism, 193-96
Patagonia, 95
patriotism: duty to settle wildness, 75, 134; national pride, 180, 205-7
PCB, 107
perceptive faculty, 53, 186
pessimism, 210
pesticides, 43, 62, 67, 72, 88, 97-104
PhyloCode, 31
picturesque, 180, 187-89
Pinchot, Gifford, 134-35, 139, 202
playgrounds, 105-6
pleasure. *See* pain
pluralizing nature, 5-7, 214-15
pollution: dumping, 114-16; economics of, 86-87, 93; Environmental Kuznets Curve, 82-83; equity, 111-17; human health, 98, 100-102, 107-9; limits to growth, 69-62; religion, 129, 137
population: aboriginal, 10; early American, 201; rapid growth, 57, 61, 138
Powell, John Wesley, 180
prairie dog vocalization, 146-47, 153
precautionary principle, 65, 211
preservation: equity, 111, 117-19; faith in technology, 66; polarizing, 1; religious motivations for, 131-32, 139. *See also* authentic experiences
pricing a human life, 111-14
pristine myth, 10-13
professions, 60, 69-72, 98, 107, 135
progressive conservation, 41, 70, 134, 139

Project Learning Tree, 203
prokaryotic, 27
property rights, 7, 128
Protestant, 130-32, 135
psychic crisis, 201
psychological health, 104-7, 202
public health: justification for parks, 105-6, 202; profession, 69; service, 98, 100
Puritan, 132, 135, 180

racism, 35. *See also* equity; speciesism
Rapture, 136-39, 212
rationality: animal rights, 166-67; human quality, 143, 150, 151
recreation: economy, 182; equity, 117; health, 105-6; popularity, 178-79, 181; religion, 135, 180. *See also* authentic experiences; leisure
recycle, 61, 67-68, 87-90
regulated forest, 44
religion, 124-40
resilience, 39-42, 45
restoration: ecological, 21, 49; psychological, 183-84
rights: animal, 158-77; members of community, 50-51. *See also* biocentric; ecocentric
risk, 65, 71, 107-9
riverhood, 171
Rockefeller, John D., 118
Roosevelt, Theodore, 78, 135, 202

scarcity, 58-61
school curricula, 202. *See also* education
Seattle (chief), 14
self-awareness, 147-48. *See also* consciousness
Seton, Ernest Thompson, 132-33
settlers: cultural icon, 200; early impacts, 13; Manifest Destiny, 74-76; pastoral, 193-95
sexism. *See* social Darwinism; speciesism
Sierra Club, 140
Silent Spring (Carson), 97-100
Smith, Adam, 76, 86
social Darwinism, 34-35, 207
soft green, 67-68
solitude, 104, 186, 204
Soto, Hernando de, 10, 75

species: compositionalism, 46–47;
economic, 78, 81; evolution, 25;
migration, 22–24; Native American
manipulation, 17; similarities among,
32, 142; taxonomy, 29–32, 171. *See
also* exotic species; extinction; invasive
species; native species; rights
speciesism, 163–64, 167
Spencer, Herbert, 220
stability. *See* balance of nature; resilience
status, 69, 135, 154, 190, 195
steer, 79
stress reduction, 104–5, 145, 155, 183
subdue nature, Genesis, 125–27
sublime, 180, 188–89
suburban: aesthetic, 190–91; land
conversion, 81; pastoralism,
196
subversive science, 54
Summers, Lawrence, 111–12
survival of the fittest, 26–29. *See also*
social Darwinism; Spencer, Herbert
sustain: sustainable development, 54, 65,
81–84, 95, 119–22; sustained yield,
43–44, 70; weak vs. strong, 120
Switzerland, animal rights, 161
symbiosis, 33

taxonomy, 30–32
technological optimists, 62–63, 65–66,
70–71
technological skeptics, 58–59, 65–66,
70–71
technology: defined, 63–64; defining
nature, 69–70; requiring
responsibility, 71–72; shaping policy,
65–67
Ten Commandments, 124, 132–33
tend and keep nature, 128
Tenochtitlán, 12
Thoreau, Henry David, 2–3, 193–200
toxic terrorism, 114–16

toxins. *See* pollution
tragedy of commons, 64
transformed lives, 204–5
Turner, Frederick Jackson, 200
Twain, Mark, 180, 186
tyranny of ecology, 55

ultimate resource, 61–63
United Nations, 88–89, 115, 140
urbanization, 98, 194, 196
U.S. Army Corps of Engineers, 181
U.S. Forest Service, 49, 99, 181
U.S. National Park Service, 99, 106, 181,
184
utilitarian, 165–66

vegetarian, 126, 162, 173–74
vervet monkeys, 146, 153
Virginia Company of London, 132

Walden Pond, 2–3, 197
Weber, Max, 135
web of life, 14, 24, 37–40, 46
weeds, 43, 56, 78, 190. *See also* exotic
species; invasive species
Wheelman magazine, 180
Wilderness Act, 181
wildness: aesthetic, 189; American
character, 75, 200–201; Henry David
Thoreau, 198–99; religion, 126, 131.
See also pristine myth
Wilson, Woodrow, 190
*Wisdom of God Manifested in the Works of
Creation, The*, 130
woods (forest), 43–44
work, 135, 179

Xerox, 93, 95

Yellowstone, 181, 185–86
YMCA, 134
Yosemite, 181, 185, 188